Ozzie.
Happy #2
lets get together
+ game on for a
Big fish
Love Grandpa!
Bill

Happy #2!
Excited Ozzie!
to explore
nature with you!
I love you,
Grandma Jan

I speak to a nearby beaver
in a soft, gentle tone of voice.

You see, I came to study you . . .

Conversation with a Beaver

and Other Journeys into Nature

by

Judith K. Berg

Judith K. Berg

Photographs and paintings by author, including cover, and within pages of *Conversation with a Beaver* and *Willow Creek Preserve*. Photographs within *Fern Ridge Excursions* by David Berg.

A portion of the proceeds from the sale of this book will be used for the preservation of fresh water habitat.

www.otterspirit.org

Book and Cover Design by Matt Gravelle.
Published by Ulyssian Publication, an imprint of Pine Orchard.
Visit us on the internet at www.pineorchard.com

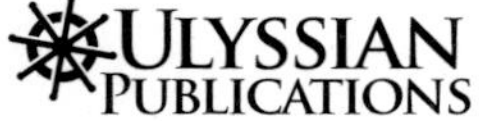

First Printing: Summer 2012
Printed in Canada.

ISBN-10: 1-930580-97-5
ISBN-13: 978-1-930580-97-8

Library of Congress Control Number: 2012939418

Printed using vegetable-based ink on recycled paper.
Manufactured with materials composed of 30% to 100% post-consumer fiber, treated without chlorine.

Dedication

I dedicate this book to my loving and supportive husband, David …

… and in Loving Memory of three special companions who journeyed with me throughout my 38 years of mammal research …

Cinder, Ben, and Rusty

Acknowledgments

In my study area at the headwaters of the Colorado River in Rocky Mountain National Park, I am grateful to the many individuals of the West Unit of the Park for their assistance, friendship, and support during my research project, particularly Jim Capps, now-retired District Naturalist, and Barb King, colleague, special friend, and inspiration.

Thank you to the partnership of multi-level government agencies and The Nature Conservancy for protecting the area known as the West Eugene Wetlands. A special thanks to The Nature Conservancy for acquiring and managing Willow Creek Preserve. With its specialized and varied habitats, it is home to a variety of flora and fauna. I am especially grateful on behalf of the mammalian species that reside or move through this protected segment of our locale. Also, I thank the Willamette Resources and Educational Network (WREN) for their important work educating people of all ages in the value of wetlands and riparian habitats to the many species of flora and fauna who reside within these environments.

I thank Carolyn Gravelle of Pine Orchard Publishing for her final edit and professional assistance in the publication of this book, and Matt Gravelle for the cover design and all aspects of the technical production of this book.

Most important of all, I thank my husband David Berg, to whom I dedicate this book. Your love, encouragement, and support throughout my professional career, and far beyond, are felt and appreciated beyond mere words. Also, thank you for your incomparable editing, suggestions, and technical work related to each story of this book and its many color photographs.

Lastly, I thank the beavers, and all of the many other faunal species, who allowed me to "look through the window" into their natural histories and their natural environments. I vow to continue to respect, protect, and preserve their worlds and their species.

—Judith K. Berg, 2012

About the Author

In 1974, when a university professor invited Judy to conduct observations on African elephant behavior in an atypical captive environment, she knew then that her contribution would be towards the preservation of endangered species of the world. Her graduate research on elephant vocalizations and associated behaviors resulted in a landmark publication on the subject.

Armed with her graduate degree, Judy became a Research Associate at the San Diego Wild Animal Park. After completing ten years' research on African elephants, she directed her studies to endangered Japanese serow, Chinese goral, okapi, and barasingha deer. Concurrent with her research, she also participated actively in local and regional nature conservation issues.

When Judy relocated to Colorado, encouragement from a number of wildlife professionals led her to research the state-endangered North American River Otter on the Colorado River in Rocky Mountain National Park and surrounds. Because beavers share the same waterways as do otters, they naturally became as much a research subject as were the otters.

From every research project that Judy conducted emerged publications in the scientific and popular literature, as well as presentations to international audiences. Although she fulfilled her life's goal, her personal drive and commitment led her to write *The Otter Spirit*.

During the long hours spent performing her research alone with the Natural World, many ideas for the short stories contained in *Weathered Wood* began to materialize. With that project completed, it was a natural progression to develop the story told in *Conversation with a Beaver*.

Judy and her husband David continue to contribute to the world of natural history during their retirement in Eugene, Oregon.

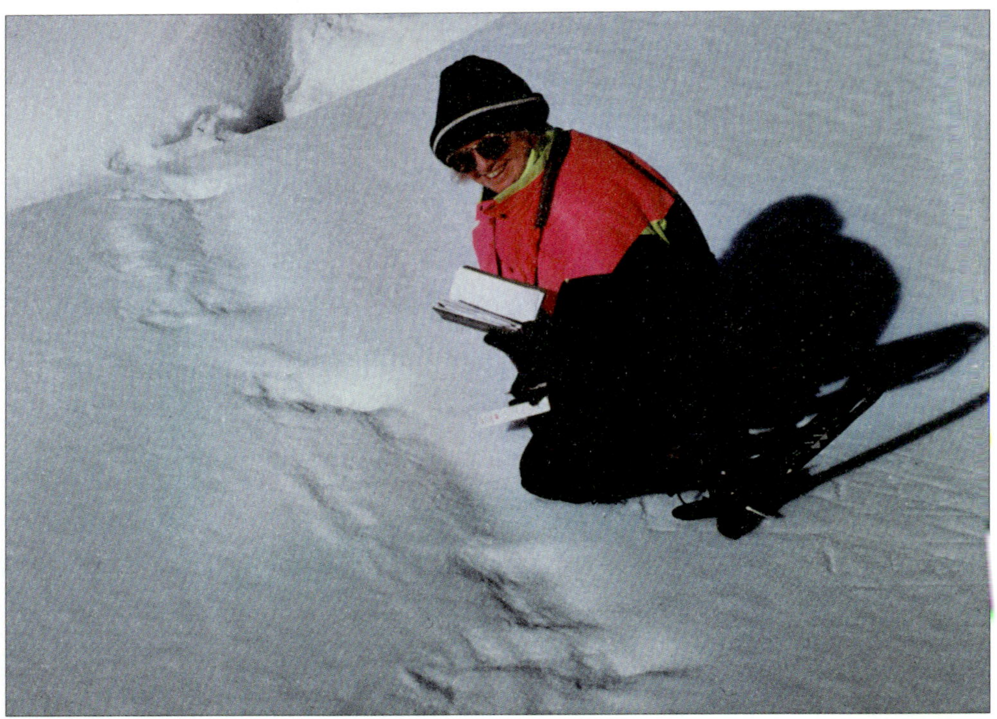

Contents

PREFACE TO
Conversation With a Beaver

During the period 1992 — 1997, I performed a study of river otters in the headwaters of the Colorado River in the Rocky Mountains. On completion of that project and its final report, I wrote my first book, *The Otter Spirit: A Natural History Story*. This was intended to be my one and only book. However, with the continuing and worsening destruction of the Natural World by *Homo sapiens*, I could not just walk away in good conscience.

Because river otters are such a mobile species, I was provided the opportunity to concurrently observe and document the behavior of another species during my project, the beaver. Fortunately, beavers are more sedentary. Beavers and river otters resided within, or moved through, many of the same locations throughout my study site's 40-mile stretch of watershed. In fact, it is shown, through my research and that of others throughout the country, that good river otter habitat is often produced by the activities of beavers.

Additionally, some of the most active periods of both species are during crepuscular (dawn and dusk)/nighttime hours. Thus, at least primarily during crepuscular hours, and more rarely on moon-lit or clear, starry nights, I spent time in beaver-created environments. This afforded me the opportunity to observe both beavers and otters. My predominant study site, as described in the journey you're about to take with me, was my favorite. Because a diverse group of other faunal species also benefited from just this one beaver family's natural masterpiece, I observed them as well.

To set the context of my study site, it was in the Kawuneeche Valley, on the west side of Rocky Mountain National Park, and nearby portions of the Arapaho National Forest, in the state of Colorado. The beaver sites mentioned were primarily along areas of the headwaters of the Colorado River, its drainages, and Monarch Lake. The terrain was mountainous, with valley habitat, ranging in elevations from about 8,000 to 9,000 feet. The riparian vegetation in my location was classified as shrub/grass, with the main shrub being willows, with lesser amounts of alder and dog birch, and many associated grasses. There were stretches of lodgepole pine, with either sparse or dense understory components. Snow began to accumulate by mid-November; the area was primarily snow-free by the end of May. Temperatures throughout the seasons during my research ranged from about -30 to +80 degrees F.

My approach to this beaver story is creative non-fiction; we will journey into their world through factual science and creative imagination. The science is based on my direct observations and the results of many beaver researchers throughout the country. Citations are contained in the "References" at the end of the story. The creativity is based on transitioning into the beaver's world through fact-based imagination. By addressing the story in this manner, I intend to intimately involve you, the reader, with beavers. I also ask you to think about how this one species contributes so much value to the Natural World and, yes, to you as well. The story provides your guidance.

Now, come and journey with me to learn about the life history and many contributions of the beaver. Then, hopefully, you will not only observe and think about their role in the web of life, but also commit to respect, protect, and preserve their lives and their home environments. In this way, you will be helping many other species as well, including your own.

Conversation With a Beaver

"Kheeeeer" resounds through the open window of my vehicle.

I recognize this as the call from the messenger bird, the beautiful red-tailed hawk. *What is it trying to communicate?*

Taking a break from my mountain journey, on the way to my beaver site, I step out of my truck to stroll into a nearby field of sunflowers. They invite me to join them in their dance of summer. We sway back and forth, their large golden-rayed blossoms ceremoniously facing toward the bright ball in the sky. As their name implies, they continually turn to face the sun throughout each day as it makes its journey from east to west.

My mind transcends to the glorious paintings of these, one of my favorite flowers, as they flowed across the canvas of artist Vincent Van Gogh. It was August 1888 when Van Gogh wrote from Arles, France, to his brother Theo about his third painting of this special flora. He decided to paint light on light: sunflowers and buds emerging from a yellow vase with a yellow background. Perhaps his major use of yellow reflected his state of mind at the time; I know this vibrant color brightens mine.

Many of this masterful artist's paintings include Nature as he saw her and as he felt. In a letter written in late September 1888 to Theo, he wrote:

". . . there are a good many more people who can do clever sketches than there are who can paint readily and can get at Nature through colour."

Even though it was over a hundred years ago that he was physically on this earth, I imagine him standing beside me and passionately painting this Nature scene. *Thank you, Vincent Van Gogh, for choosing to depict Nature through color in your many masterpieces of art and through which your depth of feelings are reflected.*

Perhaps the hawk's message to me is to move beyond just what I see in Nature and, like Van Gogh, feel her life forms, each of which is intricately woven into the web of life. I lie down to rest upon Mother Nature's warm body and close my eyes to digest this message. The hawk dips a wing towards me, catches an updraft to keep itself aloft, and soars away.

Soon, I must rise to continue my journey, transitioning from these foothills up into the Colorado Rocky Mountains, arriving in time to join my research animal of interest for their nighttime activities. The animal is the beaver, which happens to be my birth totem animal; interestingly, my related affinity color is yellow. Ah, when we again meet, I will bring them a sunflower and tell them about my favorite artist, Van Gogh. *Did he ever encounter Beaver's relatives who reside in Europe?*

* * *

Leaves, fluttering like golden butterflies, gently surround the animal that is shaking the tree that embraced them since spring. Soon, this tree that gave them substance will fall upon Mother Nature's body. The chisel-like teeth chipping

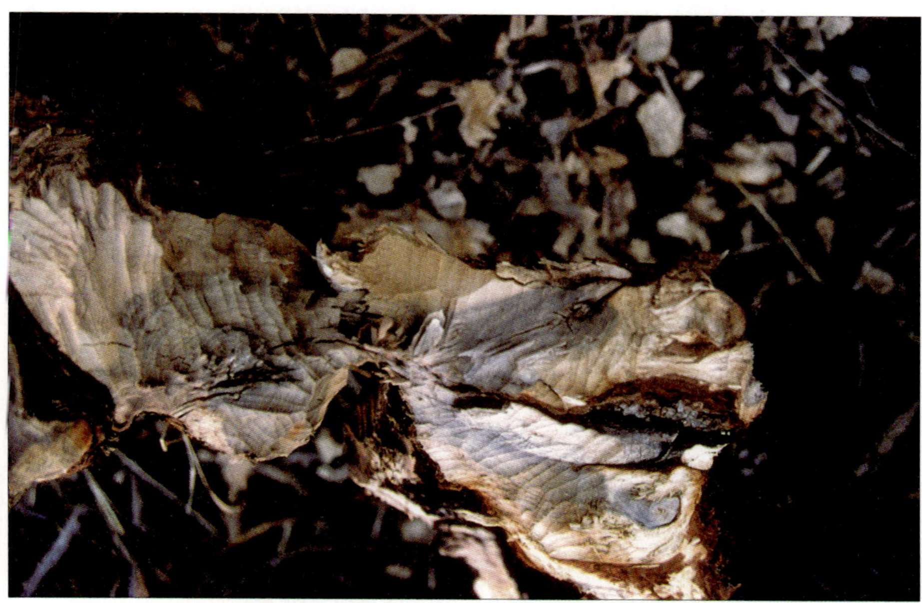

away at the tree work hard for their persistent master, a beaver. While using its lower incisors to gnaw on the wood, the beaver's upper teeth act as levers. The resultant wooden chips fall upon landed leaves of this earthen floor like tiny ships floating on the moonlit sea. Nearby waters begin to tremble in anticipation of that which will soon occur.

This scene has been performed on Mother Nature's stage over millions of years in the past and will be played out over millions of years in the future if she can continue to protect her essence and her children. Let us hope that is the case!

I arrive at dusk to quietly observe the hard-working beavers. Even though I brought my sunflower gift to them, this is not the time to disrupt the forthcoming event. So, instead, I observe and record their activities. Grasping onto the sunflower, I think about the red-tailed hawk's message and feel the essence of that which is occurring.

While the beavers are busy with their workload and feeding behaviors, darkness overtakes our world. I look up into the cosmos. A beautiful starry night presents us with a grand perspective of the vastness of our universe. *Do non-human animals ever look up at these scatterings of brilliance? If so, what do they think about them?* After all, the ancestors of many of these faunal species were on earth — and witnessed this nighttime sky — many millions of years before humans ever evolved. This is something to think about!

An urge comes over me to try to express what I am seeing in a painting. Returning to my nearby truck, I pull out a canvas and some oils (which I always keep readily available during my travels just in case I get such an inspiration). I walk back to the beavers who appear oblivious to my presence. Setting up my easel, I look to the stars and begin my artistic adventure. An unknown force bestows upon me a pulsating energy of life and starts moving my hand. The paints swirl and twist across the canvas, taking on a life of their own. I dance around my painting, stopping only to extrude some color onto this visual piece of art.

The beavers stop their activities to look in the direction of this strange individual dancing to the imagery of Nature. *How many other animal species may also be watching?* My passion continues during what seems like minutes (but is actually hours). Finally, exhaustion takes over my physical being as early dawn makes its appearance.

Is the purpose of this adventure to feel the fervor that Van Gogh must have felt while painting his masterpiece "The Starry Night"? My painting is a blur of colors, but I feel that I am able to share something very special with my favorite artist. For years, I have connected to Van Gogh through his triumphs in art by experiencing them in museums across the world. This time, however, is different. I feel his presence.

Tired, but happy, I return to my major purpose for being: studying and sharing my love for the Natural World.

"I'm sorry to have disturbed you." I speak to a nearby beaver in a soft, gentle tone of voice. "You see, I came to study you and . . . well, you saw what happened." The beaver emits a gruff, disgusted sound. "I brought you a gift," continuing my side of the conversation. "This is a sunflower."

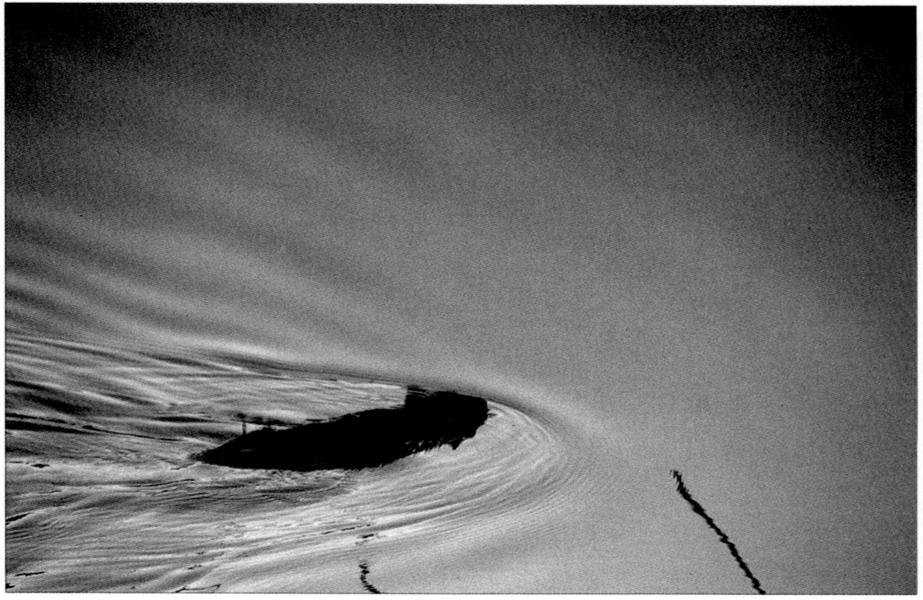

I toss the sunflower to the beaver. Looking rather astonished, the beaver accepts the gift and swims away. I feel that we can now begin our connection.

As the first light of dawn enters, the many faunal species of night are turning over the Natural World to those of day. Some species may have active periods both diurnally and nocturnally, depending on a number of variables that govern their behaviors.

The beaver dam and nearby lodge are quiet now. Their occupants had a very unusual night and, like me, must be exhausted.

What must the beaver family have thought when one of their members entered the lodge with a sunflower gripped in its jaw? (At least I assume that to be the case, since I didn't see that it had been dropped anywhere in the area.)

Beavers do not have good eyesight but instead rely more on their tactile, olfactory, and aural senses. When examining the sunflower, they would feel and smell this unusual gift. Sunflowers don't naturally occur up here in the high mountain environment, so it would have been interesting to hear their conversation. One day soon, I hope to become a part of it.

At that time, will any of them remember the strange lady with the canvas and the sunflower? I will try to explain to them what happened and about Van Gogh's role in this mystery.

Sunrise also brings forth one of Nature's musical compositions. Lovely little songbirds awaken with melodies to greet the break of day. Some songs are intended to mark territorial boundaries; some, by males, to attract females; and some may be sung just to welcome the dawning of a new day. Whatever the case, it fills the Natural World with beautiful music. A nearby song sparrow emits a solo piece. Then, in the background, other avian species sing their accompaniment. A woodpecker keeps the beat by drumming against a nearby tree. Mother Nature waves her baton over the forest as howls of the coyote and bugles from the elk enter the composition on wisps of wind breezing through the forest. This is the music of life.

* * *

Because beavers' active periods are during the crepuscular-nocturnal hours, these diurnal hours are a good time for me also to take a rest. I drive to my mountain retreat, where sleep quickly overtakes me.

After several hours, I awaken with a jolt. *Did I dream the artistic adventure?* I sit up in my cot, turn, and standing in plain sight is my painting. *Is it possible that I actually created this work?*

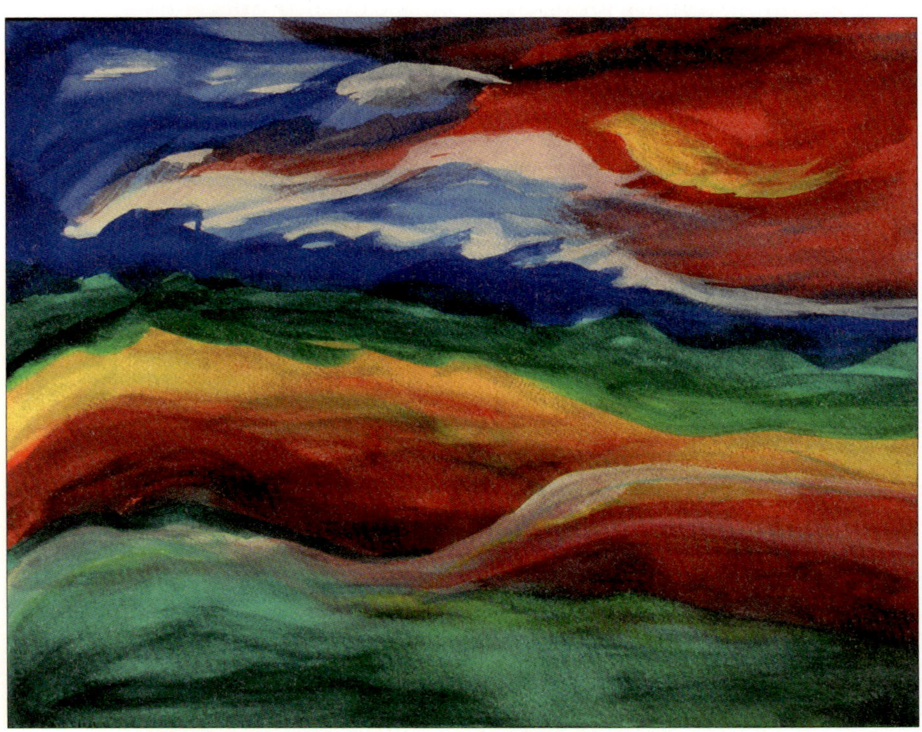

Evening is drawing closer so it's time to return to the beaver's world.

* * *

Ah, I am just in time to witness their emergence. Hiding within some vegetation, I sit quietly and observe their activities. Because I am not using radio-telemetry, maybe the crepuscular (dawn and dusk) hours will be better for my observations. After all, there probably will not be too many clear "starry nights." That may be a good thing.

Beavers are primarily monogamous, which is rare in the non-human mammalian world. Their young do require an extended period of parental care, so it may be necessary for more than one adult to successfully raise their offspring. It may also mean a division of labor between the sexes, with each adult contributing equally. We will have the opportunity to observe this behavior throughout the seasons.

Although the beaver's annual breeding season has some variance related to latitude or altitude, breeding season occurs in January or February in most areas of North America. This is the case in the Colorado Rocky Mountains. It is likely, then, that the beavers I am observing have young tucked inside one of their lodges that were born in May or June.

Young beavers are called "kits," and, even though they are born fully furred, with their eyes partially opened, and incisors erupted — thus in a precocious state (more fully developed than usual for mammalian young) — they still need considerable care. Usually two to four kits are born; and at birth weigh 8 to 22 ounces. (Adults weigh 30 to 55 pounds, occasionally even larger.) Beaver kits have a two-year development period to attain their adult behavioral repertoire. In addition to their parents, yearling siblings may also assist rearing the newest members of their family, while still continuing to develop and refine the remainder of their own adult repertoire. This will be examined as we progress on our journey.

Wishing I could become a more intimate observer of the young beavers' development, I call upon Mother Nature to lend me her hand. While using a combination of the real world and the fact-based imaginary world, this process may just be accomplished.

Through this approach, I might also share with them information about this Nature-oriented, human visual artist while they teach me about the art of beavers. After all, this species is a major craftsman. Beavers are "Nature's engineers": the architects, the contractors, and the builders of not only their homes and dams, but of their natural habitat.

Grasping onto the hand of Mother Nature, I am ready to take the plunge into the beavers' world.

* * *

Through my imagination, I enter the nursery lodge of the beaver family. As my eyes adjust to the darkness, I make out two young beavers emitting the vocalizations I previously heard above ground. Upon seeing this intruder inside their lodge, they appear frightened and emit a whining sound, which is their response to a distressing situation.

Not wanting to scare these little ones, I try to appease them, speaking in a soft, gentle voice. *"Please do not be afraid. I am your friend, not your enemy."*

Then one of the kits gets up on all four legs and approaches me while hissing. This little one is just too cute for me to feel frightened by its offensive behavior. Even though I try to hold it back, restrained laughter springs forth from my being. This is probably a new sound to these little ones.

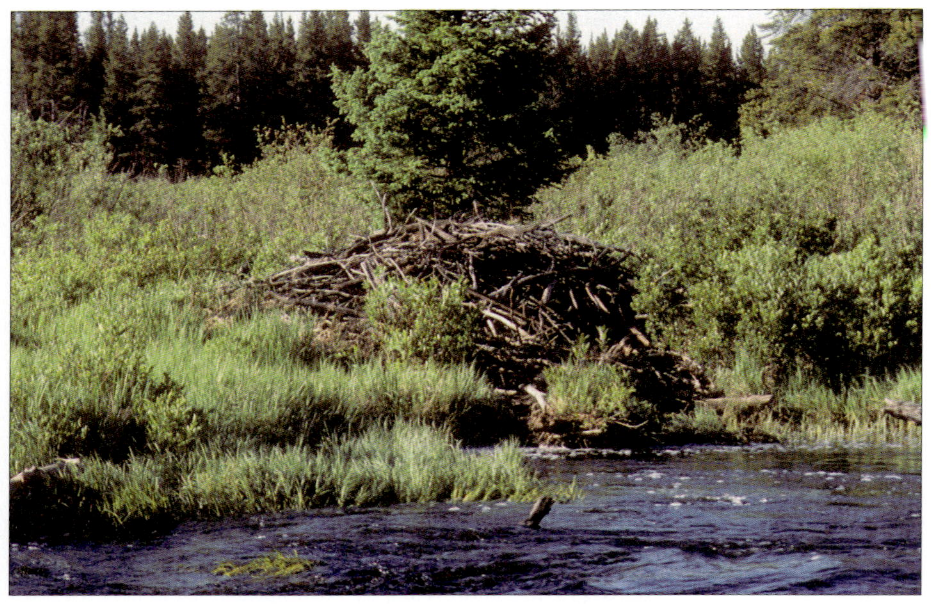

When I am moving around above ground and near their lodge, I always try to be quiet. However, on more than one occasion, when I slip into one of their beaver-created holes, my response is not a pleasant expression and my voice may sound more like their agonistic growl. I'll try to curb those sounds in the future.

As quickly as I can, I subside my laughter and sit quietly.

These kits appear to have been born in early June, so they are now four to five weeks of age. During their first month of life, they evolved inside this lodge, physically and socially developing within the confines of safety. In particular, as reported from prior detailed research, locomotion is now sufficiently developed to allow them to emerge and move around with their family within their parents' home range.

Just a few hours after birth, they attempt to swim on the surface of the water inside the entrance to their lodge. However, even after many attempts at diving into the plunge hole of the lodge, it isn't until now — at about four weeks of age – that they are able to make their first successful dive. Their initial adventure is likely prompted by a parent pushing them into the lodge's plunge hole — and with some vocal protest I may add.

They still aren't able to remain submerged until about two months of age, however. Among other factors, they're just too buoyant! During this same time period, they advance to keeping their fur water-repellent — a behavioral prerequisite to staying submerged. An element of this behavior is the development and refinement of self-grooming, which they perform by combing their fur with their claws and their teeth. Since this activity requires good coordination and balance, it also develops during this time period. We'll investigate this further in succeeding weeks.

Although suckling may continue for 10 weeks, their mother begins to bring them small tender leafy branches for their experimentation after being nursed for two weeks. These serve as toys in addition to serious food items. After all, play is an important part of the developmental process of most mammalian species.

I pick up one of the small branches and pretend to eat it. The kits watch in amazement. It appears fresh, so my behavior is not appreciated. I am again hissed at. Since they are primarily eating solid foods now, I return their meal (and toy) to them.

* * *

Thinking I have created enough of a disturbance in the kits' lives for now, I return to the real world. I emerge into daylight, as it was nighttime in the lodge while the adults and yearlings were feeding and working.

I sit down under a nearby tree and begin to write about my adventure into the nursery lodge. A chickaree — a small member of the squirrel family also known as the Douglas squirrel — doesn't appreciate my sitting under its tree, so it begins bombarding me with pinecones. Chickarees are a territorial species; their raucous vocalizations alert the animal world to my presence.

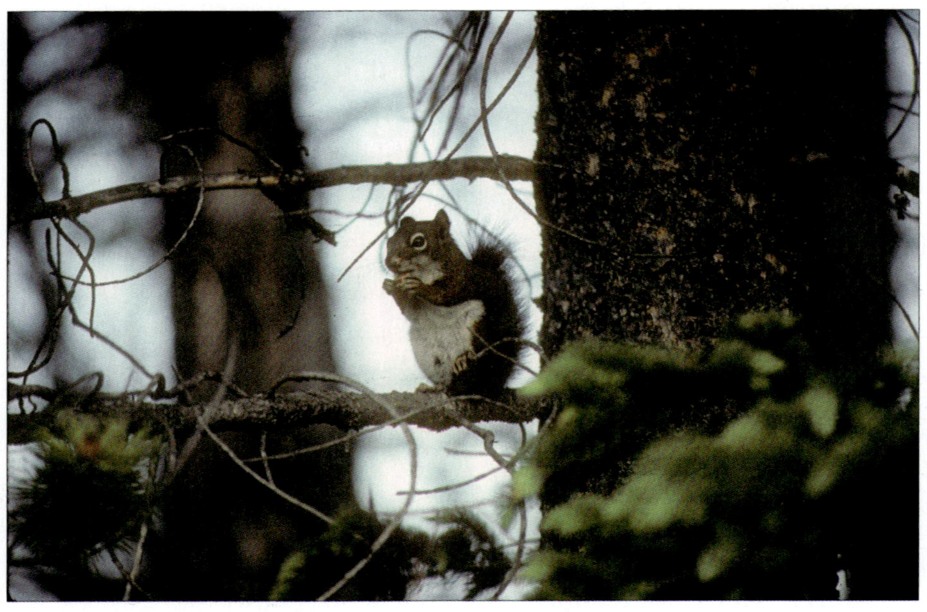

I rise and move away from the bombardier. A wisp of air blows across my cheek as Mother Nature whispers her message to me. Smiling in response, I turn and journey back to my own retreat for rest.

Because our kits are being raised within a family of adults and yearlings, many of their behavior patterns may not be fully attained until they are 10 to 12 months of age. During the next few months, the time and energy of our kits will be spent developing physically, while learning and refining different skills and socializations under the protection and tutelage of their family unit. Building and maintaining dams and lodges, for example, as well as food gathering, is left "in the hands" of more mature family members.

Some kits are not so lucky; they may be forced into a more independent lifestyle when just a few months old. Not only are they compelled into adult behaviors before their time, they are more vulnerable to predators lurking in the shadows of the Natural World.

* * *

Dusk arrives at the beaver pond . . . and so do I. We are in the phase of full moon; because the skies are clear, I'll be able to see what beaver activities are presented on tonight's stage of life. My imagination will be my only canvas. My oil paints remain in the truck.

The sky darkens and all is quiet. I sit encased within my jacket of down, leaning against a large pine tree near the beavers' lodge. Now rises the full moon, appearing larger than normal behind the forest backdrop, brightening the world below. *"Thanks, Mother Nature,"* I whisper, *"just like you expressed to me in your earlier message."*

Then, as if on cue, the beaver family emerges.

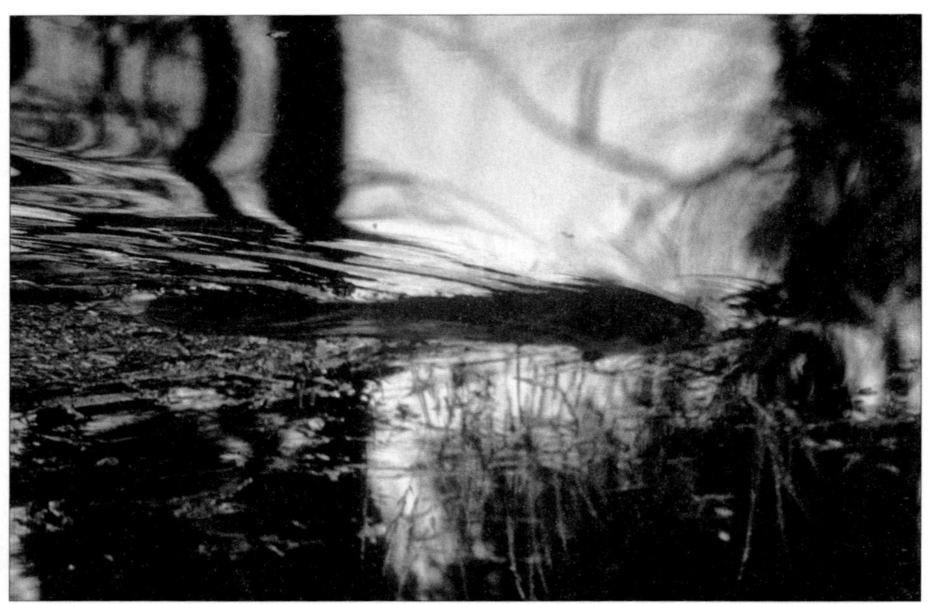

Here come the kits, bobbling around on top of the water like giant corks. The moon lights up the ripples surrounding them, reflecting shimmering beams of light back to my eyes. The kits' ability to remain submerged is still developing; complete success may still take another two to three weeks.

In the meantime, they look cute bouncing around on this liquid stage. Their antics indicate they are enjoying themselves, while remaining under the watchful eye of another family member. Although joining them would be fun, I decide it's best to just quietly observe. Deep within, though, I feel their delight.

It looks like we are all going exploring. The kits move onto land, but continue to be under the watchful eye of their yearling "babysitter." During their exploratory behavior, they remain just a few feet from the water where beavers, both young and old, seek protection. That's okay; there is enough herbaceous vegetation for them to sample and feed upon near the pond. However, during these summer months, other family members move beyond the kits' regulated distance from the lodge for a greater variety of food choices.

The moonlit night provides an opportunity for me to also observe the adults and other yearling. The yearling is moving around and exploring nearby vegetation. The adult female is feeding, which would be expected; this behavior is known to predominate while she is still lactating. I don't want to disturb her because her major role at this time is to provide energy to her young. It is good that she found some willows so she can feed on their leaves, small branches, and bark. Willows and aspen are favorite foods when available.

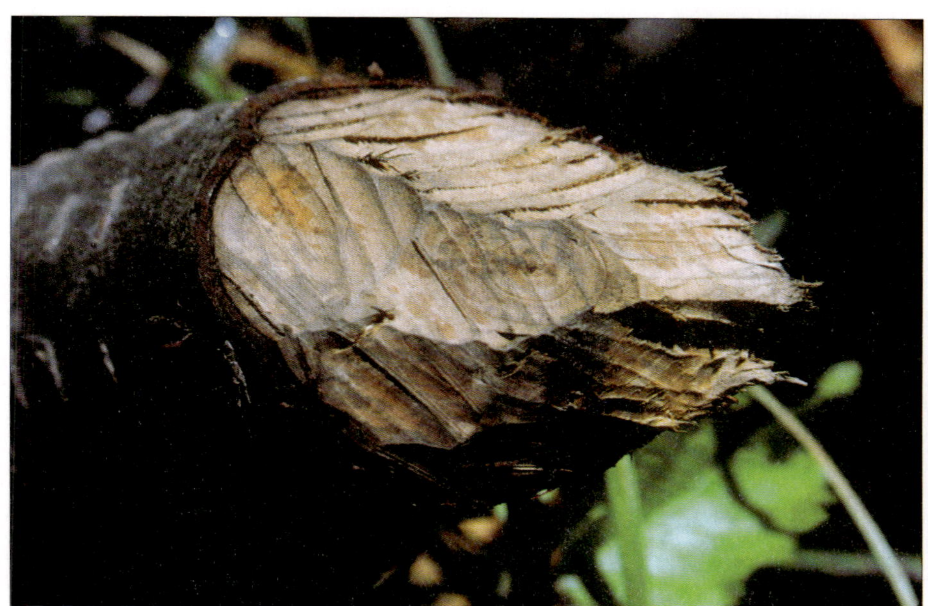

The adult male has moved farther away from the other family members, so maybe this is a good time to begin our conversation.

"Hello, remember me?" I speak softly to him. "I'm the one who brought you the sunflower gift." He stops his behavior and looks in my direction. "I'm here to learn about you, your seasonal family cycle, and, in general, as much as I can about your species," continuing my side of the conversation.

There is a moment of silence, as if the beaver is thinking about what I said. A gentle breeze brushes against our two beings. I think Mother Nature is delivering a message to each of us. Then a voice emerges from the male.

"Will you *really* tell the story of my family and my species?" There is a sense of anxious anticipation in his expression. "Will you *really*?" he repeats.

I respond emphatically, "Yes, I will."

This seems to please the beaver. "Well, we have a long road ahead, so let us begin. My major role during this period of time is primarily to provide protection and food for our kits," he speaks while gathering some small branches.

Our peaceful scene is suddenly interrupted as a large, shadowy figure enters. The male senses potential danger and runs toward the pond. Other family members respond. There is a scurry of activity.

All beavers enter the water, a tail-slap alarm is heard, and they dive out of view. They likely are moving into their lodge through its underwater entrance.

Now alone, I try to determine what created this agitation. Realistically, I feel rather insecure and wonder if I should join the beavers. No, they have enough to contend with from the outside world; disturbing them is not an option. Instead, I move closer to a large pine tree, trying to become one with my protector. While I hide within the shadows, the bark of the tree becomes my outer shell.

Nearby willows begin moving to and fro as the dark figure engulfs them.

As its shape becomes apparent in the moonlit night, I observe that it's another herbivore: the much larger moose. Although not a threat to our beavers, caution is still the best response . . . even on my part. This is a female with a youngster. Moose are unpredictable any time; but particularly, a bull in rut or a protective cow with her calf can be provocative.

The calf was probably born about the same time as our beaver kits, but much larger at 25 to 30 pounds — as much as a maturing adult beaver. Although this may sound like a big baby, a mature bull moose may weigh up to 1,800 pounds, although closer to 1,400 pounds here in the Rockies.

I usually encounter moose during the day, but maybe the moonlit night brought them out for a late-night snack. This calf would have begun following its mother at two to three weeks of age. During this time, it began feeding on vegetation, but it won't be fully weaned until about five months of age. It continues to remain with its mother until it is one-year-old and when she again gives birth.

The moose calf seems more interested in playing than feeding, as it moves around in this willowy world. In my imagination, I join the calf in play; in reality, its mother won't appreciate my gesture. Although visual acuity is not their strong point, their other senses let them know that the real me is nearby. Slowly, I maneuver to the side of the tree opposite the moose and, under cover of vegetation, move quietly away from their location.

* * *

I work my way into a nearby pine forest. Sensing movement amongst these towering giants, I become encased within their eerie world.

Outside my safety net, the real world of Nature is played out on this stage of life and death. An occasional animal sound gives me insight into transpiring activity. The interactive world of predator and prey has evolved to keep life in balance with each species playing its intended role. Although not always "pretty" to those of us caring humans, this scene has been played out since animal life began on earth, hundreds of millions of years ago.

What is not acceptable is the narcissistic attitude of too many of the more recently evolved animal species, Homo sapiens, who consider the Natural World to exist solely for their discretionary use and to be expendable to satisfy their short-term thirst for greed and consumption.

But we humans are nothing more than another species in the integrated web of life. (We've just applied our unique power of reason to meet our own selfish ends.) Like all animals on earth, we evolved from the same organism and co-exist in the same interdependent web.

We are all one.

Each species has developed and diversified to play its designated role in the web of life — most importantly, for the benefit of *all* life on earth, not just for one species. Since the evolution of Homo sapiens, Mother Nature has tried

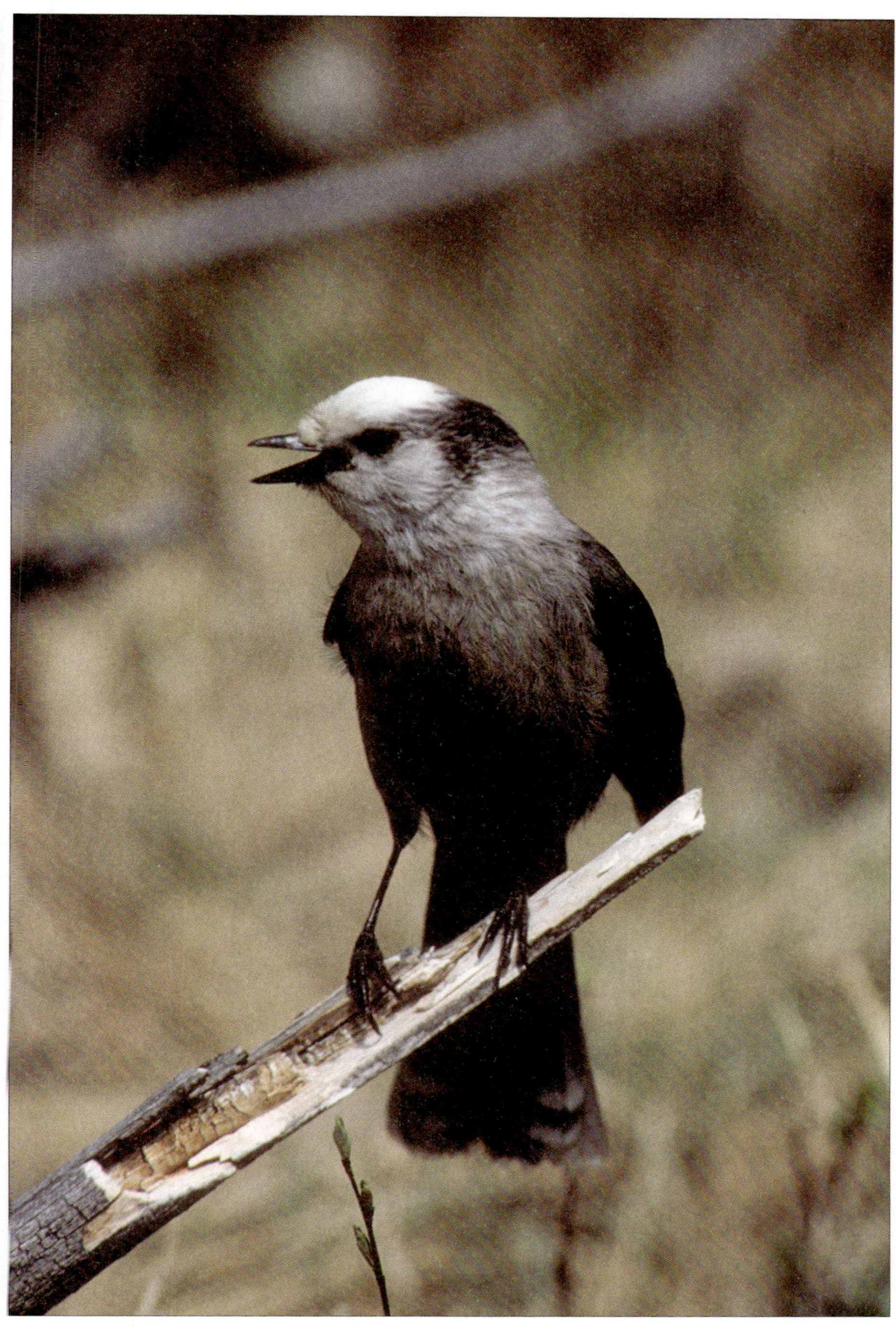

to be patient with the dominating and disruptive behavior of this one species, hoping its members would eventually realize there is nothing special about them.

Although appearing impossible, Nature and her children continue *trying* to live in peace and harmony with humans, but will it be reciprocated before it is too late?

I throw my arms around a nearby pine tree as tears begin their non-stop flow down my cheeks. I look to the cosmos and ask the Great Spirit of Nature: *"What can one caring human do to save the Natural World from those who only want to destroy her? How can one make them understand?"*

The forest suddenly becomes alive with the spirits, and, through their unison of sounds, conveys the message: *"TRY!"*

Then as quickly as this activity begins, it ends. There is incredible silence. It feels as if all life has stopped. Time stands still. I drop down onto Mother Nature's body.

When my eyes again open, first light is slowly entering our scene. Nighttime has once more turned over the stage of life to daytime. Nature's diurnal children are awaiting their performances; she directs them to begin. Sounds and activity abound.

Although there is much work and study to be conducted, I need a short break to digest that which has just occurred. Heading back towards my cabin, a breeze kisses my cheek and Mother Nature whispers into my ear.

Smiling, I respond, *"Yes, I will try."*

* * *

After a short hiatus, I return to the beaver pond. From the various colors of summer flora are emerging the yellows and oranges of fall. As this new season slowly enters our world, it brings with it changes in the life cycle of the beaver.

"I have come back," I speak to the adult male.

"Oh," he responds, "I thought you got frightened away or changed your mind."

"No, to either case," I answer. "Life just got in the way. But now, I'm back to continue our journey."

"Our family's less-intensive workload a few weeks ago has now become a frenzy of activity, except for the kits. They do try to participate and will soon learn that practice makes perfect," speaks the adult male while moving to their beaver dam.

One of the kits tries to copy him by carrying mud towards the dam. Oops! It drops before the kit arrives.

"My youngster still has a few months before reaching the advanced behavior of its yearling siblings. I have heard that some young beavers are not as lucky as mine to be brought up in a family unit. They may begin to carry out some building activities before their first winter of life," the male continues. "I'm glad our family unit of older members is able to provide the kits with the protection and opportunity to grow and develop more naturally. Watch my yearlings carry mud and branches with their front paws. And even in a bipedal gait when necessary."

"It sounds like you have much to be proud of," I respond.

Observing the beavers at work on their dams provides the opportunity to learn about the construction of these structures and the many benefits created by them.

"Since you live in a habitat with moving water, I see you have created a very large dam behind which is your beautiful pond," I observe to the beaver. "But why did you go through all that work?"

The male stops his movements and looks at me. "I am a wetland conservationist and an engineer," he replies proudly. "Also, there is an unexplainable force within me, such that when I hear running water, I feel stimulated to dam it. So I developed this large structure for those reasons and for others that I will soon explain."

"Oh!" I respond and then continue, "this dam you built is not only a fine piece of large-scale construction, but it's also a work of art."

"You mean like your human artist Van Gogh?" he replies.

"Ah, you remember," I answer in a warmly manner. "Yes, you are Mother Nature's craftsman just as Van Gogh is the human's artist. Actually, Van Gogh once stated that to understand art, one should walk in Nature. He felt that when artists love and understand her, they in turn teach others to see her. So, you see, there is an even stronger connection between you two," I share. "Well now, let's return to learning more about how and why you build dams."

"My mate and I, with help from our yearlings, built a smaller dam upstream and this larger one downstream from our lodges. This way, we hold the water level constant in this pond where our lodges are built," he shares. "So the entrances into our lodges are kept underwater for our protection from predators. You saw our family dive out of view when we thought there was potential danger. Then we all entered our lodge."

"Yes, I, too, felt concern for all of you," I reply. "Even though it turned out the animal wasn't a predator, your family's behavior certainly seemed to be the proper response."

"Thanks," he said. "Now I will continue, since you asked about our dams. The basic construction materials of the dam are varying cuts of wood, rocks, grasses, and mud. I'll teach you more details later. See the nearby water channels? Over time, we beavers dig those out to float logs and branches for our building supplies and winter food cache. In fact, it won't be too much longer before we begin storing branches under water for our cold weather food supply. Would you like to join me while I work on some dam maintenance?"

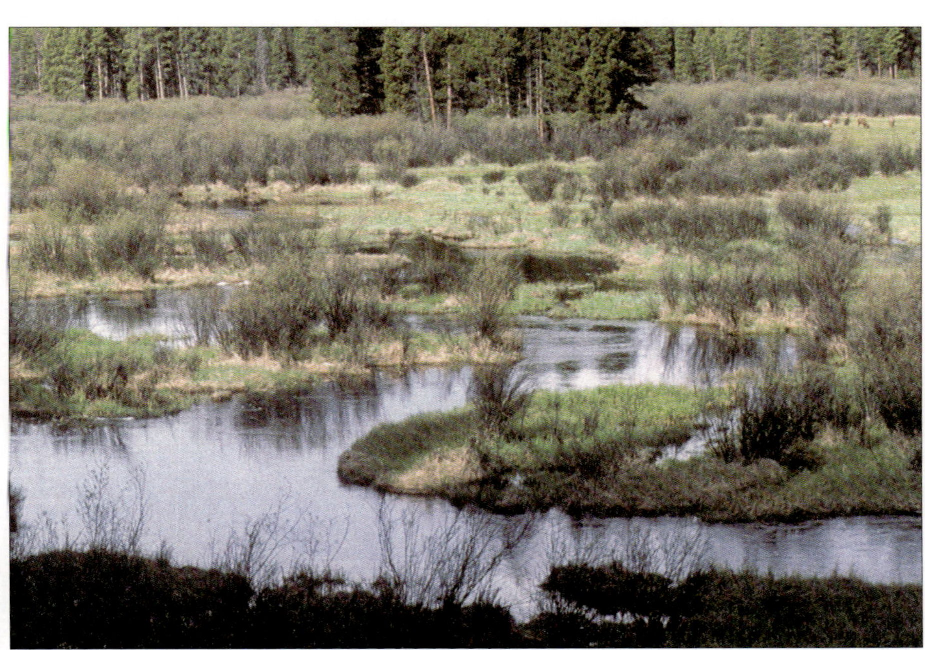

"Oh, yes!" I reply enthusiastically. "Maybe I can even help. I'll carry a branch in my mouth like you do."

"That may not be too easy for you," he responds. "I evolved with structural specializations that allow me to open my mouth when carrying branches while submerged without taking any water into my lungs. For example, watch while I close my lips behind my incisors. This allows me to gnaw when underwater. Can you do that?"

I put a small branch into my mouth then dive into the pond. Soon I am gasping for air while choking on my water intake. Sitting on the bank, I reply, "No, I guess not."

"That's okay," he responds. "While we are on this subject, let me tell you some of my other related underwater adaptations. My eyes have a nictitating membrane to keep out water. My nostrils and ear openings are valvular, so they close when I submerge."

"Oh," I interject, "another semi-aquatic animal, the river otter, also has those same adaptations."

"Ugh!" snorts the beaver. "Why do they have to enter our conversation?"

"Sorry, I don't mean to upset you," I quickly respond.

The beaver further expresses his feelings. "Well, it's just that otters have entered my lodge. Even when I was inside! These uninvited guests said they wanted a 'sleepover.' On other occasions, they moved into another one of my lodges. They said an abandoned lodge was theirs for the taking. But I did intend to use it again in the future. Oh, yeah, and one time a female moved into one of my 'abandoned' lodges with two youngsters. They even used my pond to learn to fish and swim. Can you imagine their gall?"

"In any of those cases, they probably didn't stay very long since they are usually on the move," I respond.

"Well, that's true. They are a mobile species." He softens. "Actually, young otters are rather cute and very playful. I guess they have to take time to learn and grow up just like my kits."

"This may not make you feel any better, but, from my research up here in the Rockies and other projects throughout the country, good otter habitat is often equated with the work of beavers," I explain. "As you earlier stated, you are a wetland conservationist. Your dam building and other lifestyle activities create a habitat for not only your own families, but also the families of many other species who don't have your superior craftsman skills."

"Well, since you put it that way, I do see other wildlife using the results of our family unit's labor. I guess that is another benefit to wetlands from my species' adaptations and evolution over time," the beaver acquiesced.

"You and I will continue to explore your species' importance in the web of life as our time together continues," I respond.

"Okay," he relents. "Now, if you would like, I can briefly explain about the construction of our large dam, since it seems to excite you so much."

"Oh, please do," I encourage.

"First, our family's adults and yearlings lay branches and limbs in the water across this side stream of the river. The larger ends of our wood pieces are placed facing upstream to catch any debris floating in the water to add to our dam. Then we work hard pushing mud, rocks, and vegetation into it from the bottom. Once this work is completed, we add more branches to increase its size and to help stabilize it. When we are satisfied with our 'masterpiece,' from that point on our main duty is maintaining it." He finishes, looking rather pleased with the craftsmanship.

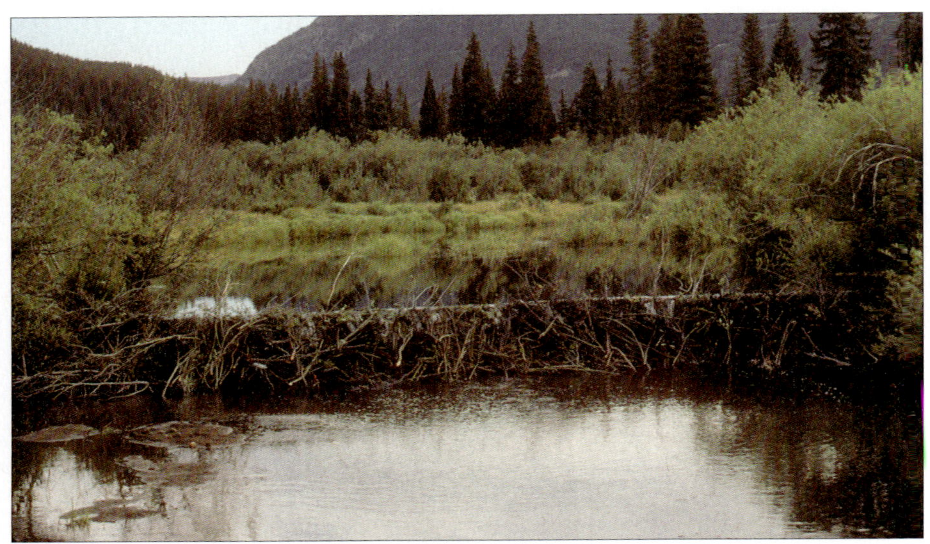

"That sounds like a lot of work," I express in a state of wonder. "I can barely imagine the labor your family has to exert, while also having to deal with outside forces. Congratulations!"

"Glad you are so impressed," he states. "Now I must return to closely inspect the dam. My mate spends quite a lot of time ensuring it is properly maintained since she is no longer lactating. But this is a behavior all family members must invest in when necessary. Good-bye for now," Beaver expresses before diving out of view.

As darkness enters our world, I decide to return to my cabin for a rest. But tomorrow, during daylight hours, I will examine other locations with beaver dams from different families in this segment of the Rockies. After all, I don't want to engage in too much conversation when there is work to be done before the next season arrives.

* * *

Emerging in the early hours of dawn, I drive to another favorite location with a beaver pond. Many hours were spent here during my research on river otters. I settle in amongst some vegetation just as an animal swims past me. Reality check . . . it's a beaver! "Hello," I say.

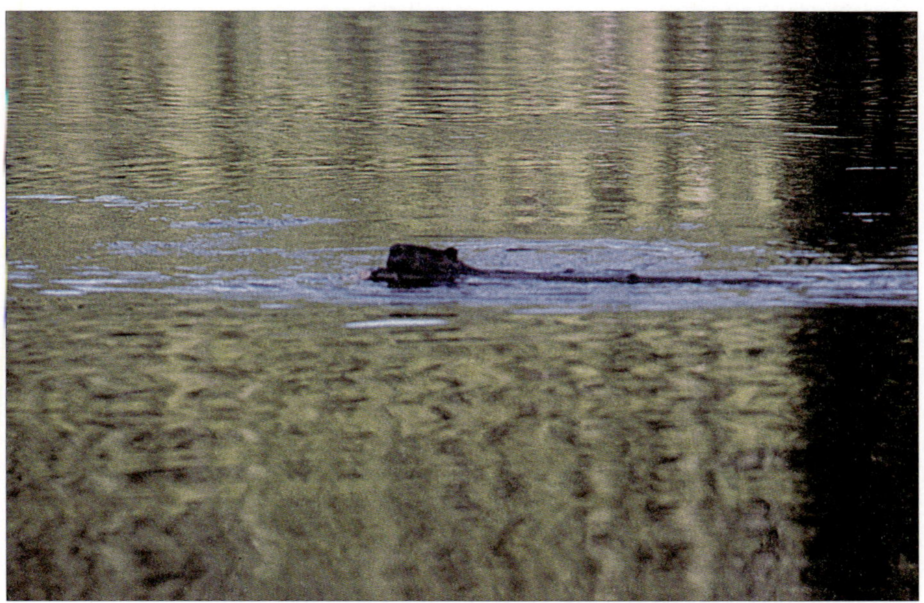

The beaver looks at me, slaps its tail against the water, and dives. I sit quietly. The animal reappears, grasps the branch it had previously dropped, and again dives. Knowing it can stay underwater for up to fifteen minutes, I wait, hoping it will again reappear. Oh, well, no such luck;

it probably swam into the nearby lodge to peacefully dine on its breakfast.

It is an evolutionary adaptation that beavers are able to remain underwater for such a long period of time. While they are submerged, the flow of blood increases through their head and decreases through their muscles so that a greater amount of finite oxygen goes to the brain.

Even though the beaver did not reappear, I wait to see what other species will present themselves in this lovely pond of quiet waters. Mother Nature's beauty is reflected in this mirror, smooth to perfection. The towering pines peek down at their own image of loveliness. An osprey calls from the top of a tall tree, announcing to the world its power and majesty while looking for a fish meal. Merganser ducks swim across the pond searching for food, but disrupting the smooth waters with an occasional bath and frolic. Two kingfishers chase each other before finally alighting on a branch. Each looks into these clear waters for a small fish or other aquatic life that move swiftly under the surface while themselves searching for a meal. A duckling swims past me. Is it alone? In the distance, I hear the falls cascading from the beaver dam as the waters work their way towards a large lake.

Before leaving, I take another look towards the large beaver lodge just in time to see a different semi-aquatic species swimming towards it. This animal is the smaller muskrat, weighing in at about 4 pounds. It swims to its house of matted vegetation next to the beaver lodge and disappears. I remember reading about an incident where a muskrat actually built its home right into the wall of an occupied beaver lodge. *Could this muskrat have done the same?*

It's time for me to move onward, but what a wonderful natural experience. There were four different avian species and two mammal species seen using this one pond — all within a few minutes of time. Just think of all the species I *didn't* see that rely on this liquid creation, including those that reside below the surface.

* * *

Continuing my beaver quest, I drive to a different location. Walking along this waterway, I discover a succession of seven small dams. Each one reaches from one side of the creek to the other. But why are there seven?

They are located about a half mile downstream from the headwaters of this creek that winds its way to the Colorado River. This beaver system contains an island lodge and another one farther along on the bank. Island lodges provide additional security for beavers from predators by surrounding their homes with deep waters. I have seen coyote tracks in this location; undoubtedly, they have, too.

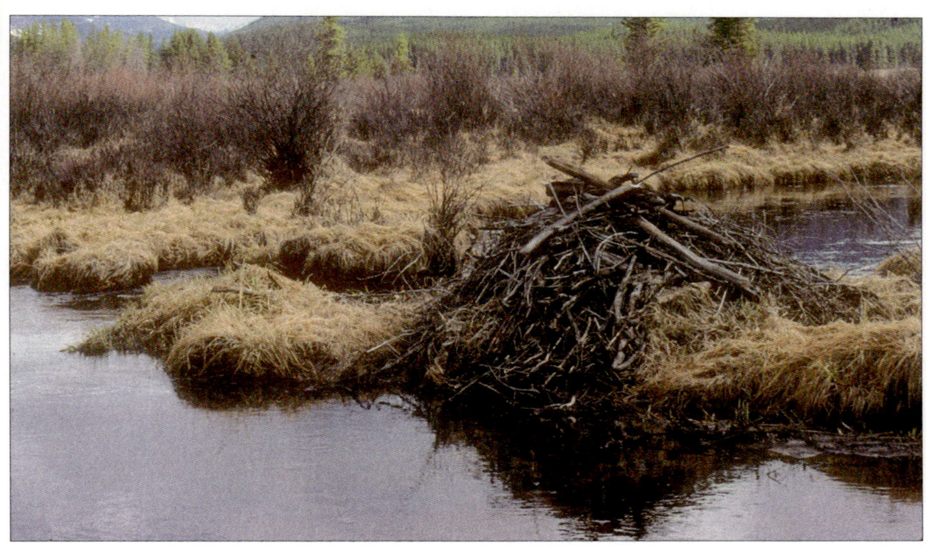

While photographing the dams, a large bull moose enters my viewfinder. He emerges from the tall willows where he has probably been foraging. There is no going-back for me at this time. The thicket of willows, the waterway, and the moose block one side of my retreat; a steep mountain slope blocks the other.

"Well, it is a nice day for a stroll," I say to myself while continuing along the creek, but constantly looking over my shoulder.

Whoops! I walk into a soggy, muddy area and begin sinking down within the muck to the tops of my knee-high boots. I certainly can't run since I can barely walk. Slowly turning around, there is no moose in sight. That's a relief.

I grab onto some vegetation and pull myself up to drier land, followed by a slow-paced progression back towards the area from whence I came. Mr. Moose is still there.

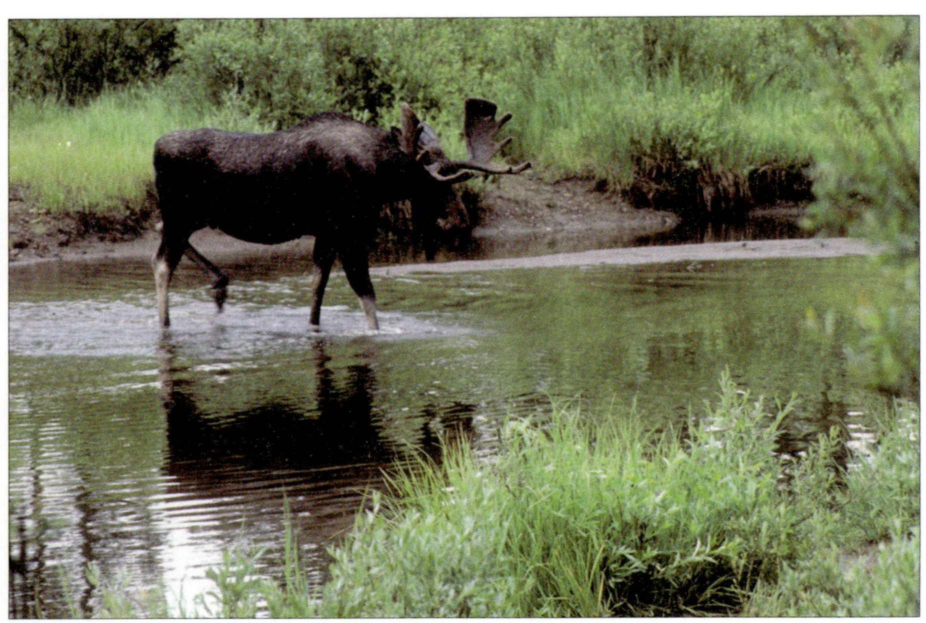

Since moose are entering their rutting period, their unpredictability is heightened. This bull is a mature male with a large rack of antlers about 5 feet across. (That's my total height!) Their species has been clocked running at speeds up to 35 miles an hour, so I try to maintain a good flight distance from him.

Hopefully, he won't feel threatened by me. *Ha, that's a laugh since I'm the one who feels threatened!* But, realistically, this is the wildlife's world and not mine.

After an hour or so, Mr. Moose leaves the area and so do I . . . while constantly looking over my shoulder with each step I take.

Exhausted, I return to my cabin where sleep arrives instantaneously. Soon, I enter a dream that interrupts my peaceful bliss. I feel myself running and running. Awakening with a start, I look around but all is quiet. *Was the moose chasing me?*

Sleep again overtakes my tired physical being and remains restful through the late afternoon.

* * *

I return to the pond of my beaver-family study site just as the colors are changing from the more intense rays of the sun to the subtle hues of evening. The violet blue waters refract the adjacent landscape into artistic shapes and colors. Then, as if on cue, the family emerges to the top of the pond.

Waters dance around the beavers as they frolic within its flow. Breezes bring forth applause from the surrounding vegetation and ripples of approval in the pond's waters. Both fauna and flora become involved in this performance on Mother Nature's stage. After a short duration of time, the beavers come up onto land to begin their nighttime activities.

"Hi, I'm back," I express with a renewed enthusiasm.

"Nice to see you," the male beaver responds. "Did you learn any more about our species' dams?"

"Only that which I observed, since no other beaver conversed with me," I respond. Then I proceeded to tell him about my experiences and the location with the seven dams.

"That's interesting. But since we don't count, that family must have built what was necessary for the same purpose that we build our dams," he stated authoritatively.

"Oh, I also saw another semi-aquatic animal that I have on occasion seen using your pond. The smaller muskrat," I share.

"Yes, I call his species 'our guests'," he responds. "They do not compete with us for food since they feed on almost any vegetative matter, and even some non-vegetative matter, unlike us. Also, because of our damming work, lots of sub-aquatic vegetation grows that we share with them, especially during leaner times of winter. So, our two species co-exist in the same space."

"Again, another animal benefits from your activities," I state. "Your ponds provide a ready-made habitat for them. Not only do you assist them with their food requirements, but also provide a place for them to build their home and raise their families."

"That's true," he responds. "Now I must feed. You may come with me as long as you don't interfere. My kits are in the process of learning about different food items that occur within the confines of our home range. There is one now trying to imitate its mother," the male points out.

"Although more selective than the muskrat, you still feed on a variety of vegetative species. In fact, one researcher referred to you as a 'choosy generalist herbivore'," I state.

"Interesting," responds the beaver. "Actually, our favorite foods are aspen and willows. But here, where my mate and I have our home range, aspen do not occur."

"Aspen are found in some other locations of this area of the Rockies, but your area has, instead, a good abundance of willows, thanks in part to you," I state authoritatively. "Through my library research in the scientific journals, I discovered that your species may be an important mechanism for willow growth. When you cut and do not eat some of their stems, they become embedded in the moist soil and generate new growth. Also, in late fall, when you are cutting willow stems for your winter food cache, these plants are

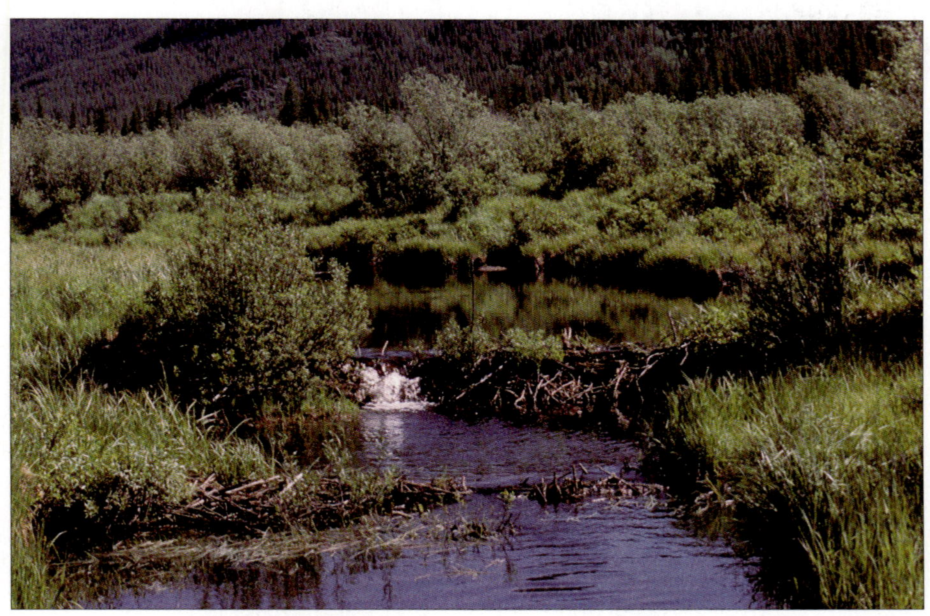

dormant. Come spring, they respond positively by increasing new stem production. It is really more complicated than that, but those are two constructive results related to your behaviors. However, the positive effects of your work can be negated when hungry elk take advantage of these new green sprouts. But, they, too, need food — especially when coming out of the difficult winter season. Actually, I haven't seen many elk in your particular home range, but there are a lot of willows."

"Thanks for letting me know what the researchers are finding out about us," Beaver expresses in a contemplative state. "Back to our feeding behavior, we have sampled other trees in the area, using our sense of smell and taste more than vision to determine what is good for us. Our selection also involves the size of the trees, distance to reach them, nutritional value, and palatability. In addition to eating bark of specific trees, we feed on leaves, buds, twigs, roots, acorns, and even some fruits. As I told you earlier, we also feed on aquatic vegetation along with our muskrat guests. During summer, we feed on various grasses as well. So, I guess you could refer to our species as a 'choosy generalist herbivore.'"

"I have one other thing to share with you before you feed for the night." I state, "In one experiment in a different segment of our country, a researcher painted the extracts of some of your potential predators' feces on aspen trees. The results of various trials indicated that those beavers ranked predators by most often avoiding aspens with feces from coyote, followed by lynx, and least often, river otters."

"We don't have lynx in our location, but we do have coyotes that we definitely try to avoid. We have discussed river otters and they have never been a threat to our family group, including our kits. Of course, they do like the habitat and homes we provide for them when they maneuver through our world. Well, I have to feed before conducting my maintenance work," he states as he moves away leaving me to sit, contemplate, and observe.

* * *

This is a very busy time for the beavers as they begin their preparations in earnest for the upcoming winter season.

I watch as members of the family carry branches from willows, then lay them down near the bank in preparation for their food cache. Eventually, these branches will be anchored to the bottom of their pond and near their lodge. These elements of Nature's flora become an important source of food when ice coats their waters. Adults and yearlings both engage in this endeavor. One of the kits even brings a small branch to lie upon this heap of future nourishment.

In addition to dam maintenance and storing their winter food cache, they also have lodge maintenance duties. Their bank lodge may begin with a tunnel dug into a steep slope with an entrance underwater. This basic construction is then enhanced with logs and branches and an internal wood floor above water level. Then it is excavated from within, with one or two entrances into the water.

After this construction is completed, they cover it with mud, debris, and aquatic plants that are brought up from the bottom of their pond. Beavers pat these building materials into place with their forepaws to seal it, except for a

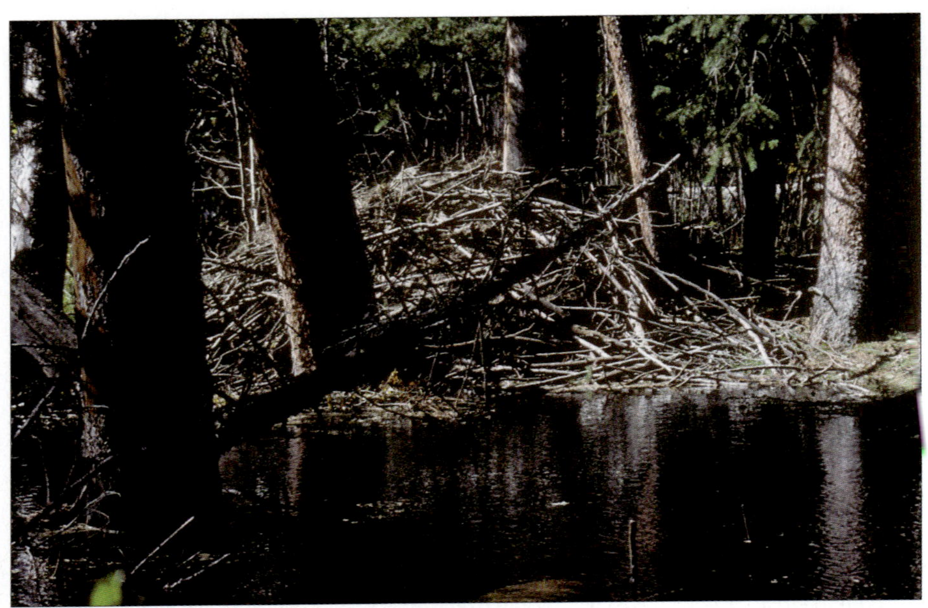

loosely-fitted dome of boughs that acts as an air vent, while still offering protection from severe weather and predators. Once the basic structure is in place, they must maintain it; this means coating the lodge with mud and debris. During winter months, this mixture freezes, providing the inhabitants with an impenetrable fortress against predators and some insulation as well.

Fall is a very busy time of year, during which the phrase "busy as a beaver" definitely applies.

We will learn more about this species' winter adaptations when we enter their world during the upcoming season. At that time, I will also share with the beavers themselves information about their species' evolution and acquaint them with a couple of Native American legends: one that relates to beavers and another to their potential coyote predator. We have much to look forward to.

In the meantime, a rest is needed.

* * *

I return in the late afternoon of a beautiful October day and in a different location to enjoy another species as it is about to perform on the open-meadow stage. The animal is the harem-structured elk, which is well into its mating season (rut).

In addition to tall pines surrounding this golden stage of grasses are trees of oranges and yellows that Mother Nature has been busily painting for the props of this natural production. Music is provided by the haunting melody of bugles from bull elk, rising in a crescendo of sounds as the performance begins.

The clashes of antlers from mature bulls also resound throughout this play. A large, dominant, well-racked male raises his head as he struts across the stage. He is definitely impressive.

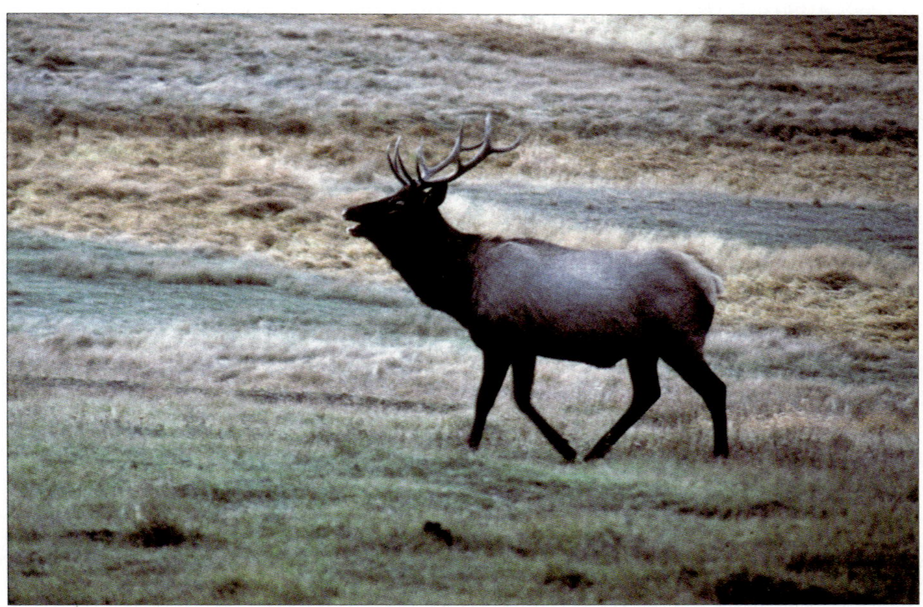

His main activity at this time of year — like all other large mature males — is to guard and hold a harem of cows and their calves. Not necessarily an easy task. I once observed a bull trying to keep a harem of 30 cows and calves from straying outside his intended boundary. In some cases, there may be even more. Also, these bulls must ward off other male challengers through a combination of displays and, as we hear, occasional head-on clashes. There seems to be little opportunity for the males to feed during this season. Of course, the reason for all of this energy expenditure is to mate and pass on their genes to another generation.

I never tire of watching this play on Mother Nature's stage of life.

I wonder if beavers ever watch this performance from their lodge on the nearby bank where the river flows alongside this meadow. I previously observed river otters moving through during this natural drama. Since both the groupings and sizes of beavers and otters are much smaller than elk, I wonder what they think of the sights and sounds of so many large actors on this special stage.

An orange glow created by the setting sun spreads across the land. Darkness will soon enter our world, but the play continues as if there is no ending.

Another species of deer, which we have briefly met, is also engaged in rutting behavior — the more solitary moose. This species adds its own melodies to this symphony of Nature. Their contribution begins with the low moaning calls of moose cows, rising to the bellowing crescendo of aroused bulls. Their mating system is a one-on-one endeavor.

While looking for other beaver dams, I recently came upon a female in her bedding area. I also inadvertently

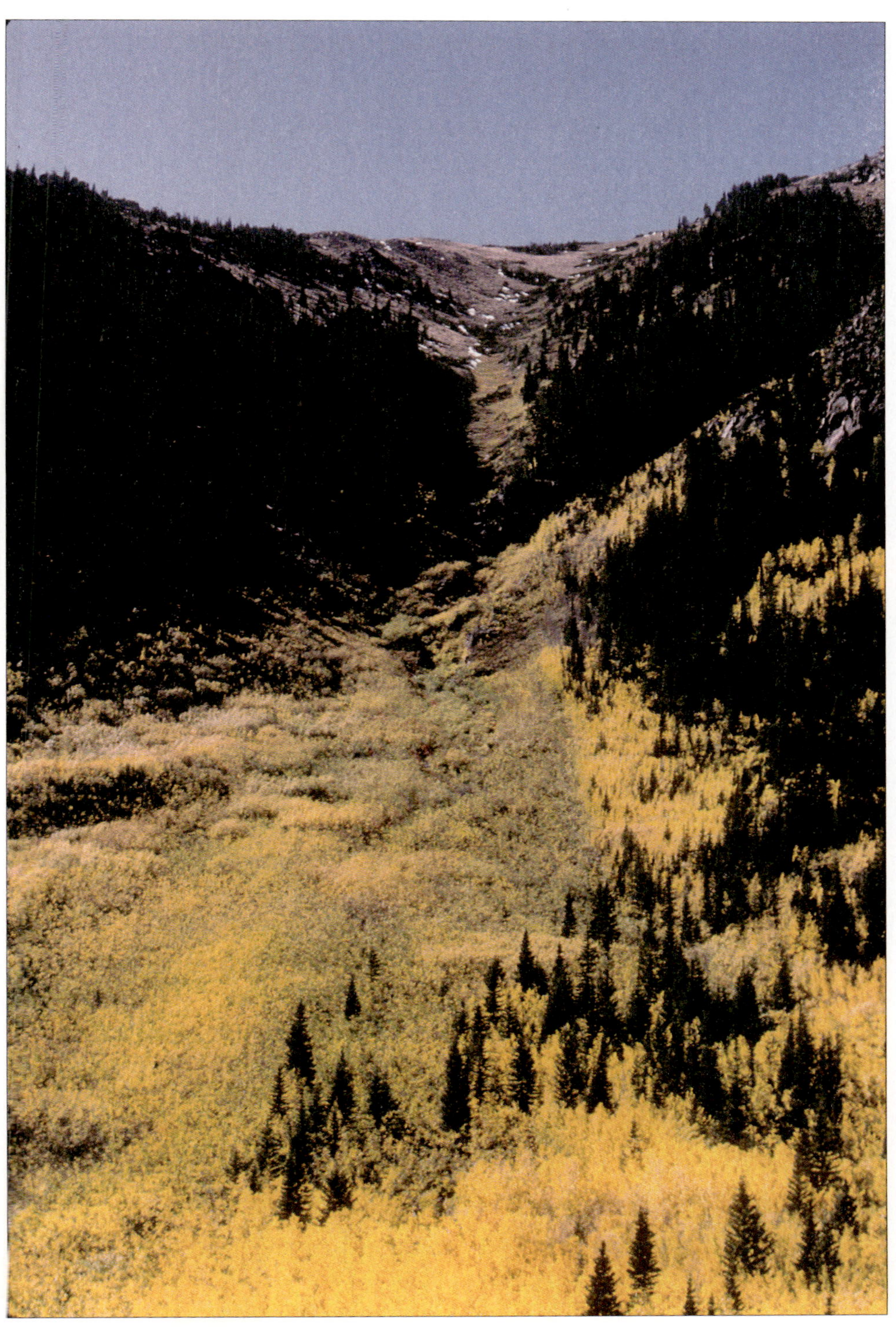

encountered a bull moose in rut while he was working his way to this cow's "bedroom." For me, this was an extremely uncomfortable situation; but fortunately, he seemed to have only one thing in mind. I responded by running away faster than my legs could carry me!

Once the two moose come together, their mating behavior will take place over about a one-week period, after which the male will move on and search for another receptive female.

Fall is an exciting and busy season for fauna in this section of the Rockies. I feel their energy and enjoy their lives through my observations and imagination.

* * *

Please join me as I return to our beaver family's pond and investigate the family's cache of winter food that remains near the waters' edge. Some of the branches have already been cemented into the mud at the bottom of their pond. This current cache appears fresh.

Another significant feature of beaver dams is that the increased water depth effected by the dam prevents the pond from freezing to the bottom in winter, thus allowing the family to swim from the lodge to their winter storehouse of food under the ice. Evolution and adaptations continue to amaze us.

Picking up one of their branches, I hold it close to my heart and feel the beaver's spirit emerge into my being.

I think about the fact that this species neared extinction in our country by the early 1900s from the European invasion into North America to trap them for their thick coat of fur. These trappers first destroyed most beavers in the

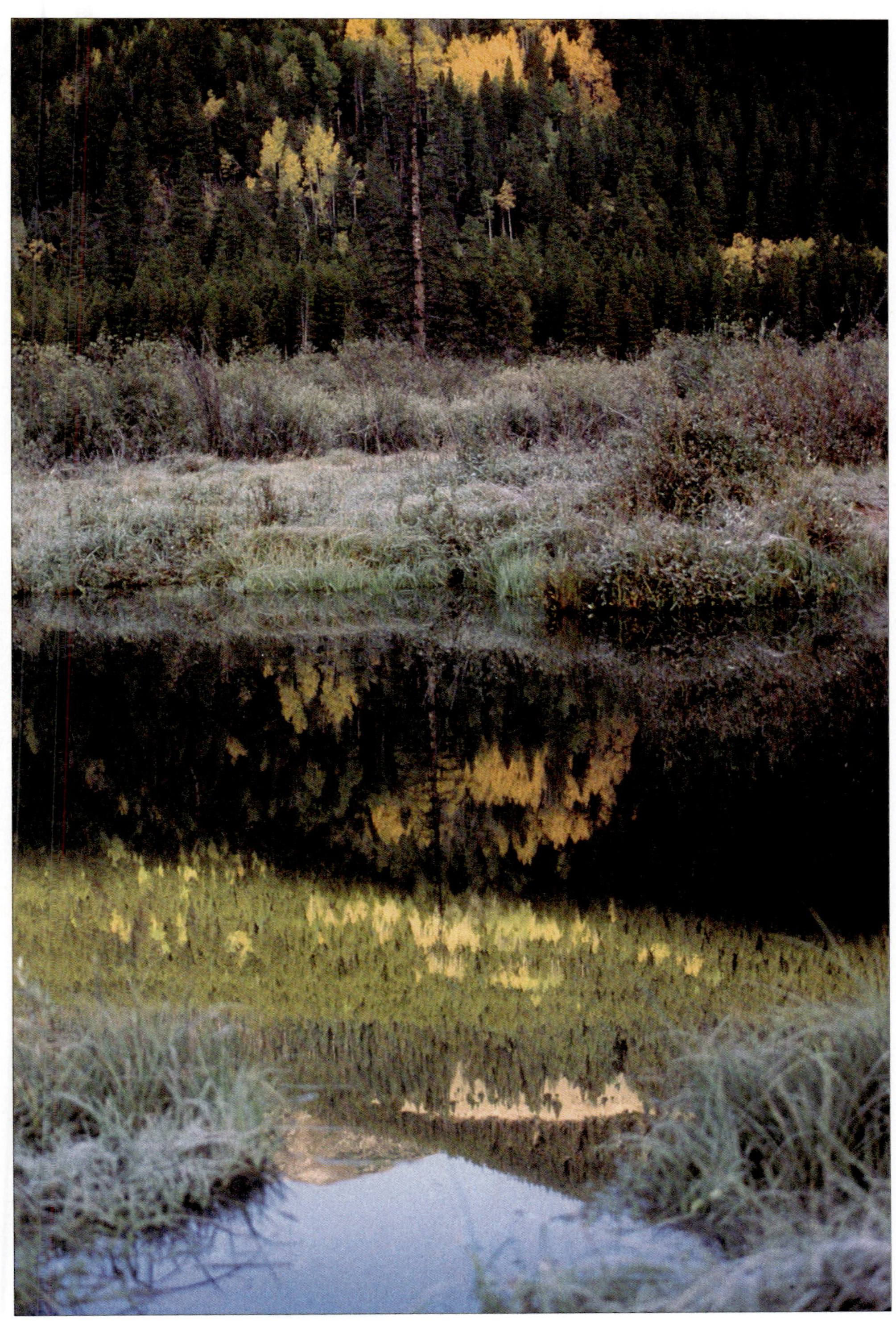

East by the late 1700s, and then continued their intense hunting while moving westward. The first wave of trappers was followed by an onslaught of European settlers. Otters and other fur-bearing animals were also taken to near extinction just so their fur could be used for human adornment and monetary greed. For the beaver in particular, which is large and easy to trap, only scattered, greatly reduced populations still existed by the early 1900s. Fortunately, through state and federal regulations and re-introduction programs, this species again exists throughout much of its former range, but in far reduced numbers from those prior to the invasion.

A cool breeze blows across these waters, sending dancing shadows across its rippled surface. The sun darts in and out from behind the clouds providing a mystical feeling that the spirits are sending another message. I gambol around the pond to embrace Nature's spirits in a dance of reverence for the Natural World.

Then I look to the skies as three red-tailed hawks are seen flying above me in layers of the currents. What a beautiful and spiritual sight to behold! One hawk lands atop a tall pine and vocalizes a message.

"I think I understand what you are *trying* to convey to me," I express while looking toward the majestic bird. "There is more to learn and more adventures to come with you as one of my guiding forces."

"Kheeeeer," the bird responds.

The beaver family appears for their crepuscular/nighttime activities. The male asks, "I wonder, are you doing another painting? I saw you dancing around the pond."

"No," I laughingly respond. "The spirits were guiding me. You see, I was just emerging from feelings of sorrow for the demise of your ancestors to joy for the existence of all current beaver families. At least that is the case in those areas of our country where your species is protected from human destruction of the Natural World and human greed."

"I'm not sure what you mean by destruction and greed since we help to build and maintain wetland habitats that are essential for so many species, including humans," asserts Beaver.

"This isn't simple to explain. It's senseless and I don't understand it either," I express. "But there are humans among us, who, rather than appreciate your contribution to the web of life, would sacrifice you and other species of fauna and flora for their own personal short-term economic gratification." During a flow of tears, I continue. "So let's just say for now that we live in the moment of this protected area while you and your family prepare for winter."

Darkness falls over beaver pond and with it comes the first sprinklings of white feathers of down — snow. *Work hard, Beavers, because Nature is sending her message to you.*

Other mammal species that don't migrate or hibernate must also accelerate their activities in preparation for winter. Chickarees are building caches of cones and nuts under large conifers in their territories. This food supply can be accessed when snows cover the ground by the animal's tunneling behavior. The slow-moving porcupine feeds on readily available evergreen needles and the cambium layer and inner bark of trees during winter. They reach their feeding trees through their well-worn paths in the snow.

Carnivores also continue their activity throughout the winter in search of available prey. Coyote, fox, and bobcat are seen digging in the snow to reach small species of voles and mice. Some of the larger carnivores, such as mountain lions, can find and take down larger prey, but also feed on similar small prey out of necessity. River otters, too, do not hibernate but continue to swim under the ice-covered waters for species of fish. They enter the waters through holes in the ice, through beaver bank dens (which have underwater entrances), and along segments of water that arc around bends in the river and often do not freeze over. "Where there is a will, there is a way" and active otters will find a way, as I discovered from my research here in the Rockies.

During our winter travels, we will investigate the deer species to see how they handle Nature's blanket of white. The road ahead will allow us to discover faunal adaptability during this often difficult season in the Rockies, and particularly for our beaver family.

"I'll return tomorrow," I yell out to the beaver. "I did not prepare properly for this current winter surprise." The intensity of the downy flakes now encases me in a coat of white.

"Hey, you look like a moving snowman," sings out Beaver. "We have nice warm double-layered coats of fur to protect us. But, I admit, we don't look forward to the period when ice will cover our pond. We hope to have some more preparation time before that happens. So, good-bye for now. We have work to do."

Trudge…trudge…trudge — over the now snow-covered trail back to my truck. Along the way, I open my mouth to feel these virgin flakes upon my tongue. Internally, I feel at peace within the quiet calm of this segment of our world.

* * *

Late afternoon of the next day, I return to the beaver pond more appropriately dressed in my warm winter clothing. The visuals of these first snows of winter are so very beautiful. Everywhere I turn, I see a light coating of virgin white as the feathers of down continue to flutter across the landscape.

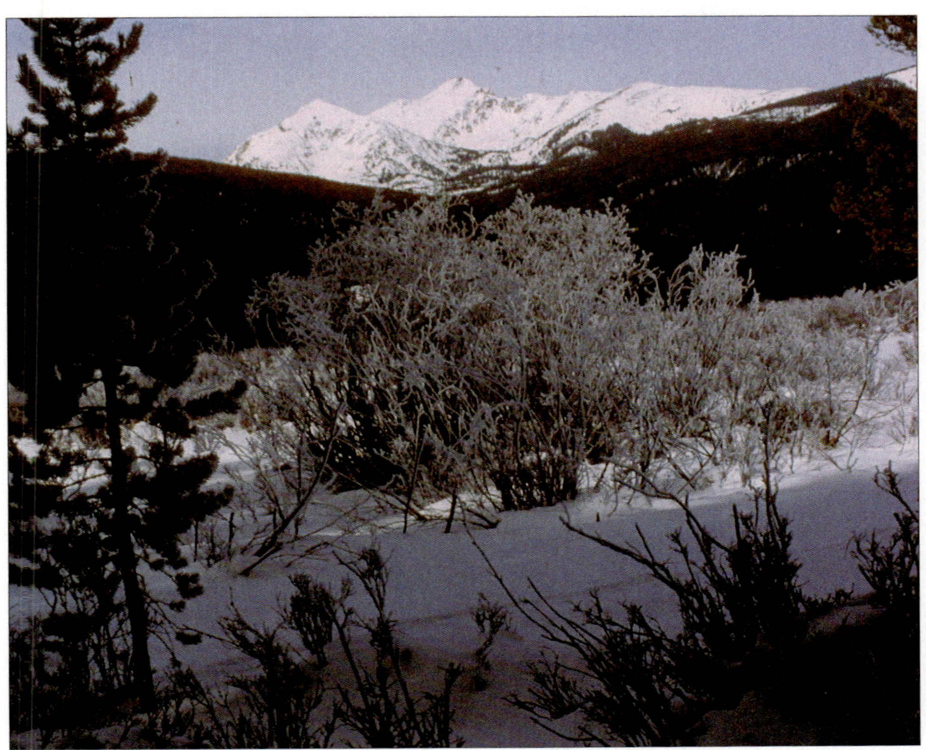

Have you ever heard the sound of silence? It feels like time is standing still.

A big splash heralds the arrival of the beaver family for their nocturnal activities as they prepare to pull branches from the shore to be cemented to the bottom of their pond.

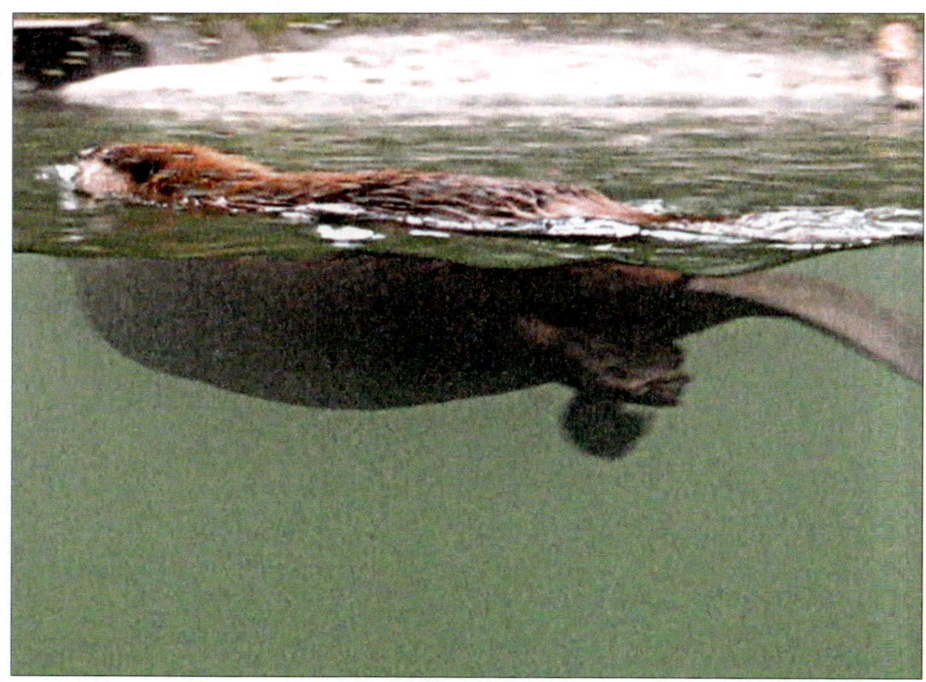

Beavers are powerful swimmers and great divers. In fact, the bone density of their limbs, particularly hind limbs, is within the range of those densities recorded for fully aquatic mammals. Also, the digits of their large hind feet are webbed. Think about flippers that people sometimes attach to their feet when swimming. Beavers, too, use their "flippers," or hind feet, when swimming, while their broad flat muscular tail is used for steering and diving. When a beaver's nose hits the water, blood rushes to its brain and heart; these vital organs would otherwise be damaged by an oxygen shortage. Another physiological adaptation of beavers is that their heart rate slows when diving; that, in turn, conserves oxygen and allows them to remain submerged up to fifteen minutes.

I stick my hand into the water; it feels very cold. *How do beavers stand this?*

The semi-aquatic fur-bearing animals (beavers, mink, muskrat, and otters) adapt to cold waters by losing heat to the water mainly through their foot pads, nose, and any other bare surfaces; in the beaver, their tail is also used for heat exchange. Most of these species' bodies are insulated by a layer of air that is entrapped by their thick double-layered fur. Their dense under-fur acts like our thermal underwear and is protected by an outer layer of long course guard hairs.

When beavers are out of the water (on land or inside their lodges), they need to keep their fur dry, waterproofed, and thermo-regulated by conducting grooming activities. Self-grooming is necessary to maintain the insulating air layer between their hair and skin. When we move inside the lodge during winter, we will learn more about their grooming behavior.

The remaining light reveals bits of floral color peeking out from their white coats of down. Branches of deciduous trees continue to hold onto a few leaves of gold and orange, shaking off flakes of snow during light breezes. Most leaves have already fallen to the ground, beginning their recycling process to nourish the soils of life. Needles of pines seem to welcome the moisture, or are they just showing off their beautiful contrast of dark green encased within this purity of white?

The falling snow does not deter the beavers, who continue to conduct their activities of feeding and working on land, while also shaking off snow that attempts to cling to their bodies. Shaking is one of their necessary self-grooming behaviors to eliminate moisture from their fur. Actually, this behavior also makes them appear to be engaging in solo dance performances on Mother Nature's stage of life.

I conjure up a rhythm to their unknown musical adventure. Standing, so able to view them without interfering in their activities, I work on their cadence. The kits seem to have their own particular beat with more shaking than the older members of the family unit. Perhaps like most mammalian young, they have some excess energy to exert. So, I concentrate on the older beavers. I imagine their behavior as a sort-of beaver dance with steps: *feed, feed, feed, then shake . . . move, turn, move, then shake*. And the beat goes on. Okay, I'm going to try this dance.

"Hey! This is fun," I sing out while whirling to the rhythm and shaking off the snow from my own outer clothing. Eventually, I fall upon Nature's blanket of white, laughing at my rather silly interpretation.

The beavers stop their activities and look in my direction. The male asks, "Are you okay?"

"Sorry," I respond. "I didn't want to disturb you, but my imagination carried me into a different dimension. In answer to your question, yes, I am just fine. Please continue with your necessary work."

* * *

The next day, I venture to the creek with seven small beaver dams. It is now the end of November. A layer of ice has begun coating these waters. The only open water occurs along the series of dams, flowing gently over these structures before moving under its frozen cover as it heads to the Colorado River.

The virgin snow on the bank reveals coyote tracks. However, this beaver family is protected — tucked safely within their cozy island lodge, which, although surrounded by frozen water, is now coated with ice, creating an impenetrable fortress.

I continue walking towards the river. The marshy area is now covered with a crusty snow; my snowshoes allow me to move over it while hugging the icy waterway. Along my route, I discover some river otter tracks and a slide. This is always a cause for excitement to me. Through my imagination, I picture the animal moving across this natural ice skating rink.

Finally reaching my destination, the river reveals a pattern of open waters interspersed with patches of ice. I sit next to one of the frozen patches to observe a species of fauna working from underneath this covering. Chips of ice are thrown out on top of the river's overlay.

Even after an hour of observing this sporadic behavior, no one appears. Whatever species created this interesting event remains an unsolved mystery.

Moving to the top of a nearby bank, I sit upon Mother Nature's coat of white to absorb the silence. After a while, an increase in wind carries with it music that appears to be from a Native American flute. A howl from a distant coyote enters this musical composition. What proceeds is interplay between flute and coyote. Soon a few mountain snowbirds chime in with their melodies of song. A woodpecker keeps the beat. What beautiful music!

I look around but see no human player. Perhaps the flute-like sounds are coming from the winds blowing through an instrument of Nature. When the winds subside, so does the music.

Silence resumes.

* * *

Returning to my vehicle, I drive past the meadow where, just a few weeks ago, the presentation of rutting elk was in full production. Now, a more-subdued, mixed herd comes into view. The herd contains about 100 males, females, and young. Most are lying down upon the snow-covered grasses; a few remain standing, pawing at the ground to reach nourishment.

Continuing along on my drive, a porcupine saunters along the roadway. I stop to take a quick photo of this seldom-seen species. It looks towards me, then continues moving to a destination only it knows.

A bit farther down the road, a coyote is also walking alongside the road. When I approach, the coyote moves up a nearby bank and pounces into the snow. It likely scents one of its small prey in hopes of getting a quick meal. When it pulls its head out from the snow, I see that it was successful.

* * *

It is now the end of December and ice cover is complete. Snowshoes are necessarily my second pair of feet. They allow me to walk over mounds of snow to reach our beaver family's pond . . . except, that is, when moving through strands of willow bushes as I break through the less densely packed snow around willow stems, followed by growling expressions of dismay.

Whoops, I hope the kits don't hear me.

Finally arriving, I am greeted to a picturesque site.

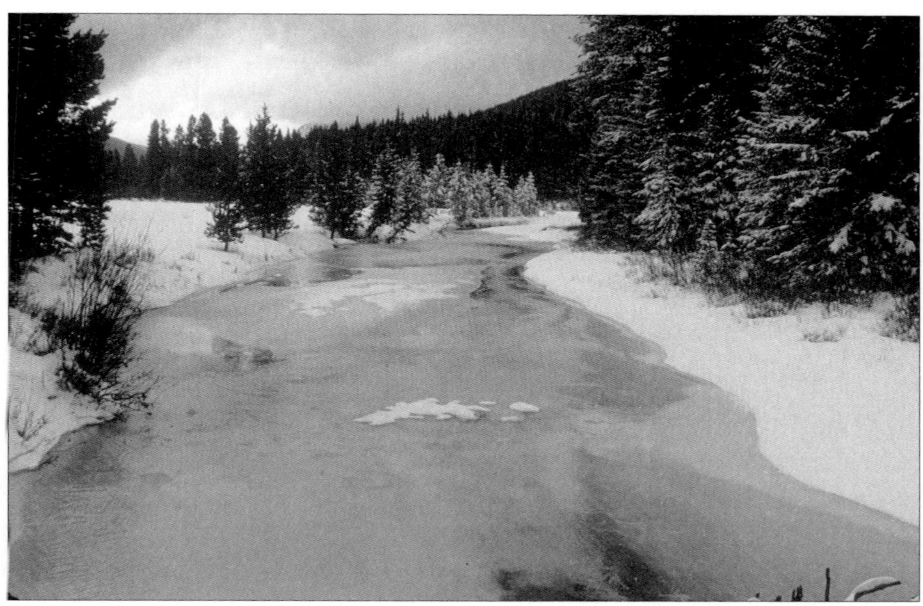

Sections of Nature's cover are translucent, so I peer through this looking-glass world to see what mysteries might be presented. How nice it would be to have x-ray vision! Because that is not the case, my imagination is my guide.

"Hello, down there," I call out. "I have come to learn about your winter adaptations."

There is no response.

For now, I just look for any recent beaver signs, such as tracks, that would indicate they came out from an opening somewhere. There are no apparent signs. That is expected, as the ice cover across the pond appears thick.

Over the past few days, we have been in a deep freeze with ambient temperatures ranging from about -20 to +10 degrees F. Since these are the respective temperatures outside and inside my cabin this morning as well, I don't blame them for not wanting to emerge, even if there were an opening in the ice. They are probably tucked within their nice cozy lodge insulated from the cold ambient temperatures. Only dedicated human researchers are silly enough to be out in this weather.

Inside, beaver lodges are both warmer and have temperatures less variable than those outside their shelter during the cold of winter. Research has shown that, even if ambient temperatures drop to -30 degrees F, the beaver lodge could be as warm as +42 degrees F. Some researchers believe the temperatures of the surrounding waters resemble those inside the lodge — at least above +32 degrees F. Although not all researchers agree on the exact internal lodge temperatures during our extreme cold, we do know they remain much warmer than those outside their dwelling.

Also, these inner temperatures are influenced by a combination of factors such as beaver activity inside the lodge, thickness of construction materials used for the walls, and Nature's helping hand of coating the outside walls with insulating snow. Whatever the case, I look forward to soon joining them.

Behavioral and physiological mechanisms have evolved in beavers that favor adaptive strategies of under-ice activity rather than above-ice activity during extended periods of subfreezing temperatures. During these cold temperatures, even if there are any areas of open water in the ice-coated beaver pond, research has shown they would not emerge to avoid suffering a significant energy deficit. Their under-ice mechanisms are what we are going to discover in the next few weeks, including how they keep their lodges so nice and warm.

I make my way to their major lodge, where I assume the family is huddled inside. Actually, huddling reduces heat loss to an absolute minimum. Although the walls of the lodge are frozen, providing good insulation for the interior, the vent located at the apex of their house remains open; this is probably due to warming-related activities by the beavers inside.

I listen for any activity that might be occurring inside the lodge. But all is quiet. Reduced activity — along with periods of dormancy — is another of their forms of energy conservation.

Sun shines upon the newly fallen snow, which glistens like precious stones spread across the landscape. Crystal chandeliers hang down from branches of pines, twinkling in response to the bright light. The surrounding mountain peaks are the walls of Nature's cathedral, spiraling towards the incredibly deep blue sky. This astonishing scene is provided to all who look and feel its beauty.

Holding out my hand to Mother Nature, I ask, "Will you once again guide me into the beaver lodge through my imagination?"

A whisper of wind blows across my cheek and soon I find myself inside the beaver's home. I rub my eyes to become used to the darkness and soon see the family huddled together, as I suspected.

"I am sorry to disturb you. But I promised to share with your family a couple Native American legends and information about your evolutionary history during this cold weather," I express.

The male beaver opens his eyes and responds, "Actually, your stories will provide a nice break for our family during this difficult time of year."

"That's good," I respond. "So now everyone get comfortable and I will tell you a story from the Tlingit tribe of Alaska called 'Beaver and Porcupine.'

"First, let me explain briefly that porcupines are animals that have sharp quills that cover their bodies, except for their face, the underside of their belly, and their tail. These quills stand erect when the animal is frightened, but otherwise lay against their body. They do not shoot out their quills, but they are easily discarded. Like you, they also feed on the bark of certain trees; but unlike you, they can climb high in the trees to reach their food source, so they are not competitive with your species. Here's how the legend goes."

> *A long time ago, Beaver and Porcupine were good friends, and Porcupine would visit Beaver's home. However, Beaver didn't like him coming inside because he left behind some of his sharp quills. So, one day, when Porcupine told Beaver that he wanted to visit his house to eat some different trees, Beaver said, "Okay, just get on my back and I will take you across the water to my lodge."*
>
> *Porcupine got on Beaver's back and Beaver swam. But instead of going to his house, Beaver*

went out into the middle of the lake and dumped Porcupine on an old stump, thinking he would sink.

This was a frightening situation for Porcupine. However, instead of giving up, he began singing his power song: "Let the water become frozen, let it become solid ice; then I can walk ashore and not sink."

The cold North Wind responded and blew in, freezing the lake, and Porcupine walked home.

Later in the year, Beaver and Porcupine again met and started playing together, so once again were friends. Then Porcupine said, "You come now. It is my turn to take you on my back for a ride." Beaver was reluctant but finally got on Porcupine's back, and Porcupine took him to the top of a very high tree and left him there.

The kits begin to whimper, probably thinking that one of their own was left at the top of a high tree and couldn't get down. I reassure them that everything is okay and hand each one a small branch I found on the floor of the lodge. All quiet down and I continue the story.

Well, after a period of time, Beaver finally got up the courage to try to walk down the tree trunk using his sharp teeth to make notches, like the rungs of a ladder. So, to this day, you can still see notches all over different kinds of trees. Beaver did this just in case he ever gets stuck up in a tree again.

The beaver family seems pleased with the ending of the story. Now, you readers, also have a mythical explanation as to why there are beaver teeth marks on some trees in their habitat.

"I will now depart so all of you can continue with your activity — or inactivity — during this difficult season," I express. "From prior research on beavers living in colder environments, when confined to a subnivean existence, beavers were found to not live on the 24-hour clock of the open-water seasons. Instead, they had a free-running activity rhythm. This behavioral adaptation reduces active periods to assist in reducing energy expenditure. It makes sense that this also applies to your family," I continue. "However, your entire family may not always be synchronized in their cycles, because activity inside the lodge helps keep the inside temperatures at a comfortable state."

"I am not sure exactly what that means," Beaver responds. "But I know we do whatever we can to keep ourselves warm and healthy."

"I will ask Mother Nature to help me with the proper timing, so I return when you are all awake. Then I will share with you some information on your species' evolution."

"That sounds good to us," Beaver responds. "The kits are already asleep, so the rest of us will join them . . . or will we?"

* * *

Returning to the nippy ambient temperatures back on the surface, I glide over the snow-covered landscape to my truck. Driving to my cabin, I think about the wildlife that remain in this montane environment and the hardships they must endure. Even though these species evolved with

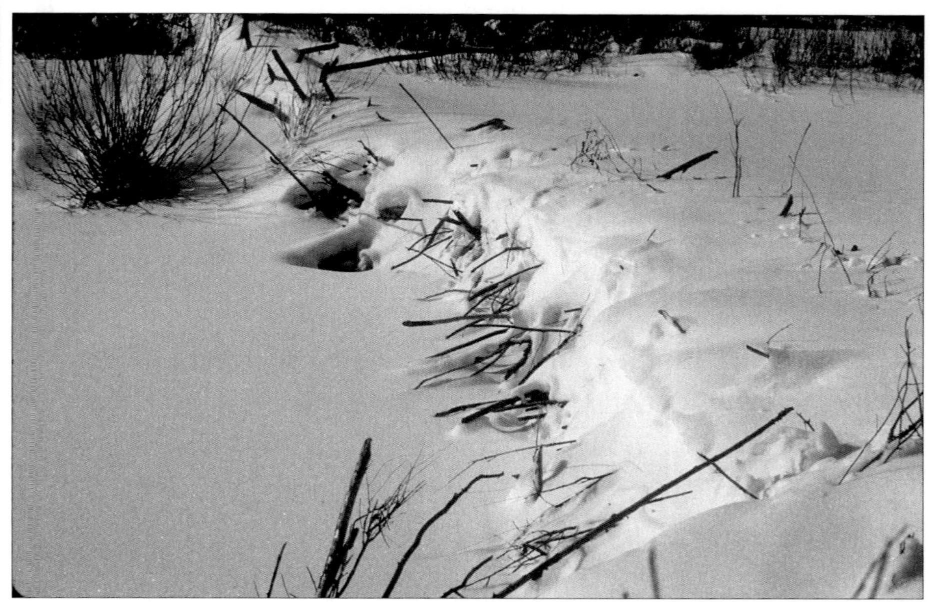

adaptations to assist them during winter, extremes of cold add stresses that threaten their survival.

Finally reaching my cabin, I enter the "freezer" and start a fire in the old wood stove. Wrapping within a blanket of down and sipping hot tea, I sit next to the stove and think about the beaver family and all that they must endure. I definitely have nothing to complain about.

I pull out a scientific journal I brought with me and read an article from research conducted on beavers' winter-feeding behavior. Among research subjects, adult beavers feed very little during winter, while yearlings and kits feed above maintenance levels. The suggested rationale is that youngsters continue their growing process throughout winter. That makes sense.

The healthy cache of food cemented into the pond near the lodge of my beaver family certainly looks like enough for all members to feed upon. However, I have not conducted an extensive analysis that would be necessary to draw such a firm conclusion.

Continuing with this article, which reinforces other research I have read, it is shown that adult beavers use their own fat reserves during winter, including fat from their tails. In fact, research performed in a northern climate finds that the fat content of beavers' tails increase from about 7% in spring and summer to over 60% in fall and early winter! That evolutionary adaptation is simply amazing.

Further, body weight and tail measurements taken before and after winter submersion below the ice reflect that emerging adults lose weight and have a decrease in the dimensions of their tails. In both cases, this is concluded to be the result of depletion in their fat reserves. In fact, it is determined that adult beavers can lose up to 34% of their initial body weight before their own maintenance systems break down.

Even though adults may feed little from their winter food cache, they and their yearlings are the beavers that bring branches back to the lodge. This division of labor avoids exposing the kits to the cold waters, since their body temperatures cool much faster than the older members. So the kits remaining within a family unit are fortunate to have altruistic parents and older siblings performing behaviors that promote their survival. This is something most pleasant to think about.

Before falling asleep, I think about the complexity that exists in the Natural World. Mysteries abound for the curious human to try to solve and for those who try to ensure that the web of life will continue for eons in the future as it has for eons in the past.

Mother Nature is *trying*.

* * *

The next day, I return to the pond and lodge location. It is cold. At least, there aren't any winds. Wow! It looks like plumes of smoke rising from the vent at the apex of the beavers' lodge.

Maybe they are burning wood in their version of a stove. Actually, it isn't smoke from a fire, but condensed vapor from their collective exhalations — an indication of activity inside the well-ventilated lodge. This may be a good time for my entrance. Mother Nature assures that to be the case.

When inside the lodge, I observe both individual and mutual grooming behavior. Remember, grooming is primarily specialized toward maintenance of the insulating air layer between a beaver's fur and skin. Mutual grooming allows the animals to help each other by reaching areas the individual has difficulty reaching itself, such as its flanks, back, and neck.

In addition to shaking, which we earlier discussed, grooming is accomplished by using their incisors or claws of their hind feet — in particular, the double claw of the second toe. However, they will also use their forepaws as a comb, stroking them through the fur with their claws deeply plunged into it.

"Hi, I'm back," I say enthusiastically. "You can continue with your grooming behaviors while I try to briefly explain what is known about your evolution. I realize this will be difficult to understand, but I will try to keep it relatively simple."

"Okay, we are ready," responds the beaver.

"On the geological time scale, your ancestors emerged in the New World during a period called the Oligocene. Through adaptive radiation from the Order Rodentia, they appeared during the early part of this period — about 36 to 30 million years ago. They were a different species, from the genus *Palaeocastor*, thought to be more terrestrial than your current species' semi-aquatic lifestyle.

"Fossils were found from one of your ancestors in Nebraska during a later time period, the early Miocene (23 to 16 million years ago). This species used burrows that were referred to as 'devil's corkscrews.' At the end of a tight corkscrew was an underground living chamber, about 8 feet below the surface of the ground. From some animal fossil remains, this species was determined to be about the size of your current-day guest, the muskrat. Paleontologists discovered preserved walls showing teeth and claw marks, determining that these animals used their incisor teeth to excavate while pushing loose dirt back with their front and hind feet."

I take a deep breath before continuing and to give the beavers a break.

"All this information you are giving us sounds very complicated," Beaver expresses.

"Yes, that it is. But we will progress into a relatively more recent time period very shortly," I respond. "One interesting note is that through progressive evolution of beaver-cheek-teeth, there has been an increase in the height of the tooth crowns (hypsodont), which is an adaptation for gnawing wood.

"Now we come to your current day genus of *Castor*, which may have migrated back and forth from Eurasia to North America across one of the land bridges. Previous fossil evidence shows that the earliest 'true' beavers on this continent were thought to have settled here about 5 million years ago or during the early Pliocene geological time period (5.3 to 3.4 million years ago). At least, that was the case until a recent discovery.

"This newest find (2011) is a pair of teeth from a beaver found in an Oregon paleontological dig, that shows to be between 7 and 7.3 million years old, or during the late Miocene geological time scale (11.2 to 5.3 million years ago). In fact, these teeth are determined to be almost identical to our living animals. This latest evidence is thought to lend credence to the time period when the two living species diverged: *Castor canadensis* of North America and *Castor fiber* of Eurasia.

"I still have another interesting piece of information to share with you before I leave," I express. "From fossil evidence during the Pleistocene geological time period (1.8 million years ago to recent), your species may have co-existed with much larger forms until about 10,000 years ago, when this larger species became extinct. This species is called the Giant Beaver, or *Castoroides*, and was three to four times as large as you! They occurred both in North America and a smaller version in Eurasia. Giant beavers lived in lakes and ponds bordered by swamps. They were unlikely to have felled trees like you do, but had short legs and large webbed feet suggesting they were powerful swimmers."

"You have given our family a lot of information to digest," Beaver speaks in a thought-provoking manner.

"I think I will go to the food cache and bring back some branches for us to digest nutritionally."

"That is very clever," I laughingly reply. "I, too, will depart and let you carry on with your foraging behavior. Next time I enter your lodge, we will have another legend, one related to your potential coyote predator that will make him look rather foolish. Good-bye for now."

* * *

Returning to my world, I feel like playing in the snow. Explaining beaver evolution has me mentally exhausted. Actually, this is good packing snow, so I soon find myself creating a beaver snow sculpture. I wonder what the family would think upon seeing this white replica of their species. Although cold and tired, I like sharing information I have learned about them and appreciate that which they share with me.

This is a good opportunity to investigate winter adaptations of some of the other species that remain in this montane environment. After all, the beavers might appreciate a break from their evolution lesson.

Trudging across the snow-covered terrain, I am drawn to signs from what appears to be a miniature river otter. On the ice-covered waterway are tracks and a slide leading into an opening. The slide is only about 3 inches wide, compared to my measured otter slides of 8-10 inches. These current signs were made by the smaller mink, which is in the same taxonomic family — Mustelidae — as the otter. During winter, they usually feed through a hole in the ice and travel underwater, digging into mud for any hibernating species such as frogs and crayfish or even a swimming fish.

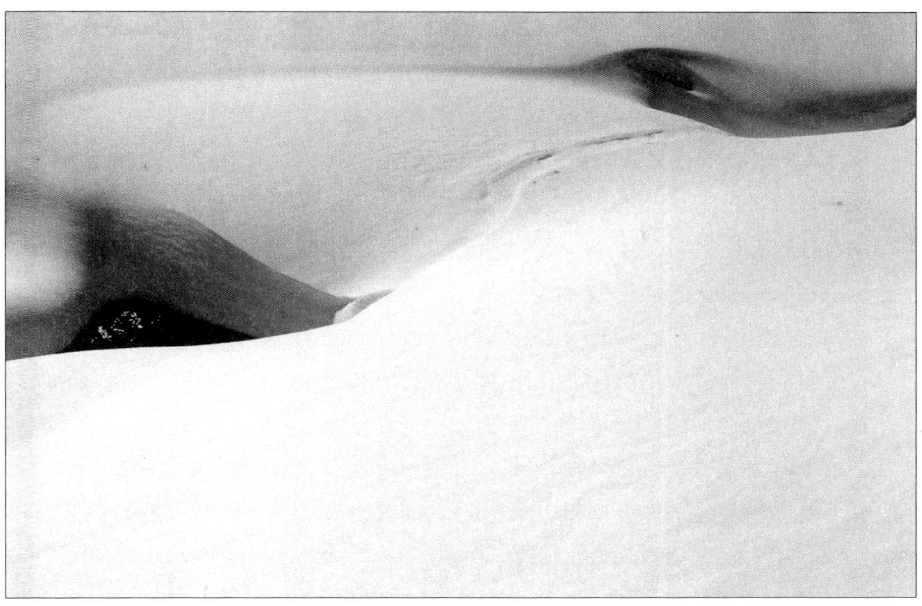

Otters and mink are both semi-aquatic, carnivorous mammals that feed on fish and crayfish. While this is the main diet of the otter, the mink is more a generalist, opportunistic predator with a diversified diet. They feed on almost any meaty creature that comes their way, unlike the otter. Because of this resource partitioning, these two species can co-exist in the same fresh water ecosystems.

A terrestrial carnivore I see on rare occasion is the red fox. Watching it pouncing in the snow bank to reach a small rodent is quite a site. Oh, there's a fox now loping along on the crest of the snow with its beautiful reddish-colored coat contrasting against the purity of white. The animal stops and tilts its head, as if listening, then pounces into the snow for an assumed vole or mouse species that are their major winter food source.

I once found a winter fox den with some tracks near the entrance. Interestingly, some otter tracks appeared just a few feet away. *Did they encounter each other?*

Our three species of deer in the Colorado Rockies are the moose, elk, and mule deer. We discussed and encountered the first two fore-mentioned animals, but not the third. I seldom see mule deer. They prefer more forested habitats, where they are well concealed. However, on occasion, when we're both trudging through the same treed locations, we surprise each other.

During winter in the Rockies, elk and mule deer usually move down the mountains to lower-elevation winter ranges. I have observed elk pawing through snow to reach any remaining forage, but apparently mule deer do not engage in this behavior. Even so, when snow depth reaches about 16 inches, elk respond by moving to their wintering ranges; however, they have been reported to move through snow at about 42 inches in depth, if necessary. Elk rely on their body fat plus good winter browse, often stripping bark from trees, especially aspen. This is a very apparent sign I have encountered.

Mule deer diet changes from summer grasses and herbs to winter twigs, buds, and bark from various trees and shrubs. Because they are smaller than elk, their movement to wintering ranges is triggered when snow depth reaches 6-12 inches. This message must be heeded because as the snows continue to pile up to about 20 inches, they can no longer exist in an area.

What about the larger moose? They are a northern animal, more adaptable to cold weather conditions. Adults cope with most snow levels because their long legs lift their chests above the surface up to depths of at least 30 inches. However, they can tolerate snow depths of about 48 inches at which movement then becomes critical. They feed on dry, woody material during winter, especially the buds and bark of deciduous trees and foliage of evergreens. This contrasts with their diet of aquatic vegetation and willows during open-water seasons. Also, during winter, moose may actually move to old growth forests at *higher* elevations; the forest canopy filters enough falling snow from reaching the ground for the moose to partake of its provided browse.

Even with all their winter adaptations, species residing in snow country have a very difficult time. Not all individuals will survive, even those at the top of the food chain.

* * *

Returning to the beaver lodge, I can tell there is activity occurring within by observing the vapors emerging from their "chimney." Once again, I call upon Mother Nature to allow me to join my special family within their nice cozy lodge.

"Hi," I sing out in a joyful manner. "I have returned to share with you another Native American legend. But this time, one related to your potential predator, the coyote."

The kits appear concerned and curl up closer to their parents.

"Oh, please don't be afraid," I speak in a soft gentle tone. "This story will make you smile at how silly Coyote sometimes behaves. This animal is an ancient mythical symbol for many Native American tribes. Even though he is often referred to as a trickster, he is also a teacher. We all learn from both his mistakes and his accomplishments.

"I will now share with you one of my favorites, which is from the Pima Tribe of southern Arizona as told by Frank Russell in 1908. It's titled 'The Bluebird and Coyote.' So you'll know, we do have mountain bluebirds during warmer weather up here in the Rockies. They may even come to feed in the nearby meadow of your habitat. Okay, now get comfortable and we will begin our story."

At one time, Bluebird was a very ugly color. However, there was a lake where no river flowed in or out; and the bird bathed in it four times every morning for four mornings. Every morning, the bird sang the following song:

"There's a blue water.
It lies there.
I went in.
I am all blue."

On the fourth morning, it shed all its feathers and came out of the lake in its bare skin! But on the fifth morning, it came out with blue feathers.

Well, Coyote had been watching the Bluebird. On that fifth morning, he said, "How is it that all your ugly color has come out and now you are blue and beautiful? You're more beautiful than anything that flies in the air. I want to be blue too."

Coyote was at that time a bright green.

"I went in four times," said the Bluebird and taught Coyote the song.

So, Coyote went in four times; and with the fifth time, he came out as blue as the little bird.

Now, he felt very proud. As he walked along, he looked on every side to see if anyone was noticing how fine and blue he was. He looked to see if his shadow was blue too, so he was not watching the road.

Presently, he ran into a stump so hard that it threw him down in the dirt, and he became dust-colored all over. And to this day, all coyotes are the color of dirt.

It looks like the beavers are smiling.

"I see what you mean about Coyote behaving in a silly manner," responded the male beaver. "We are content with our reddish-brown coats that Mother Nature has given us. She has ensured that our species coloration is what we need to adapt to our environment."

"You are correct," I say in a contemplative state. "That is a good lesson for all species, including humans. Oh, I have a question for you. While I was talking, I noticed some of you chewing on wood and some of you grinding your teeth. Did this legend upset you?"

"Actually, that is not the reason for our behavior," Beaver responded. "We gnash our teeth or chew on wood to keep our ever-growing incisors from becoming too long. If we don't do so, their increasing length will eventually prevent us from closing our mouth and feeding."

"Ah, yes, that's true," I say. "Once I saw a photograph of a beaver skull with its incisors curled around and embedded in its skull. So, please chew and grind as much as necessary to ensure nothing like that happens to you or your family."

"Thanks," Beaver expressed. "Oh, just to let you know, my companion and I are about to enter our annual mating season. She has an estrous cycle that lasts about two weeks. We will mate either in the cold waters or here inside our lodge. Then, when we are successful, in another 100 to 110 days, we hope to have new kits. They will be born at the optimal time of year here in this montane environment."

"I will be sure to give you privacy," I giggle. "I have learned a lot about your survival strategies during the extreme cold of winter. Also, your close-knit family appears content even while residing together in such a relatively small space. That says a lot for your species."

"Our adaptive strategies have evolved over time to ensure the survival of our species," Beaver speaks in an authoritative manner.

"We are entering the month of February. The early thaw will begin to occur in a few weeks," I share. "This is a good time for me to take a break from the Rockies and return again when we can meet on land within your home range. There are still some important aspects of the upcoming spring season that I hope to learn about you and your family. Good-bye for now."

"We look forward to your return," expresses Beaver.

I think all species are patiently watching for any signs of the upcoming season of spring. This is the period of rebirth for both the flora and the fauna. It is a time of jubilation.

However, I also think that winter is a season for exultation. Without the snows in our colder climates and the rains in warmer segments of our country, there would be no fresh water and, therefore, there would be no life.

As I dance across the landscape, I stop to pick up a ball of white and say, "*Thank-you*!" Then my attention is drawn towards the majestic snow-clad mountains and I hear them singing.

Listen.

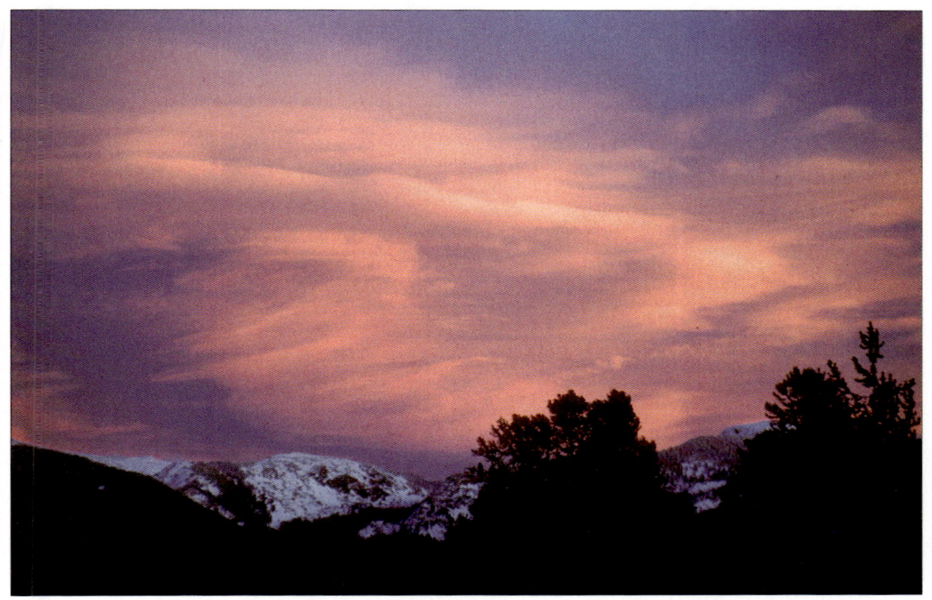

All the voices of Nature sing out to the world. *Feel their rhythm*. Mother Nature has so very many lessons to teach us. Please listen and heed them!

* * *

After a short hiatus, I drive back up into the mountains where I am greeted by lots and lots of snow. The roads are relatively clear, but the surrounding drifts of white appear like giants ready to fall upon any unsuspecting passers-by. Even though the calendar suggests it is spring, these snow-clad sculptures resist melting into their liquid form.

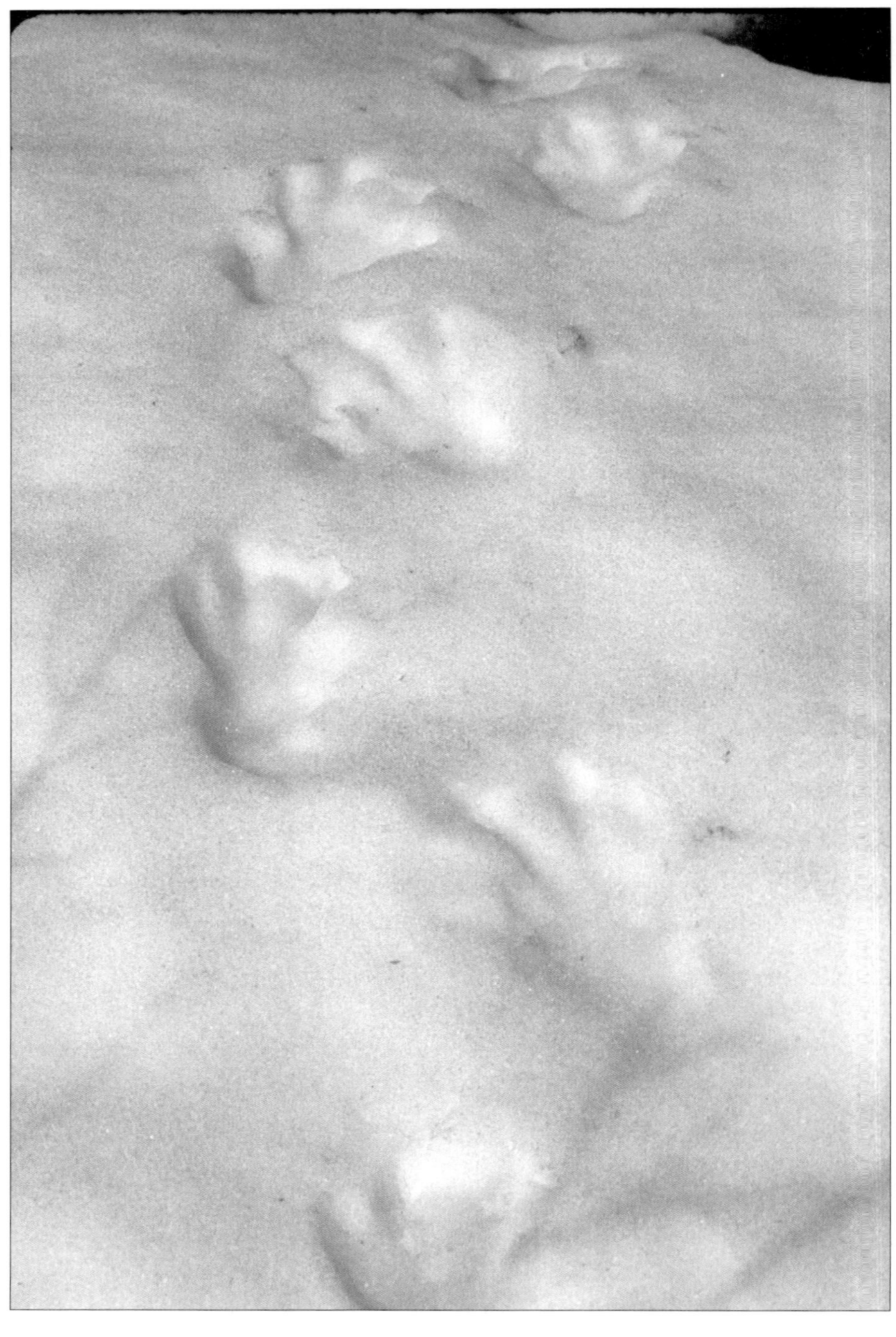

However, as mid-day temperatures reach above freezing, their endurance may be short-lived. This also means that those of us using these roadways during the crepuscular hours must be very careful. That which melts and flows during the warmer part of the day freezes in place during our still-cold nighttime temperatures.

My beaver family's complex lies at about 8,800 feet in altitude, so snow and ice continue to be the major features of the landscape. After pulling off the roadway, I put on my snowshoes and make slow progress toward their home range.

Near the pond, I discover some otter and beaver tracks which must be recent since just two days ago we had 6 to 8 inches of new snow. This coating would have covered their previously-made signs. Both sets of tracks are within a few feet of each other.

Was there an encounter?

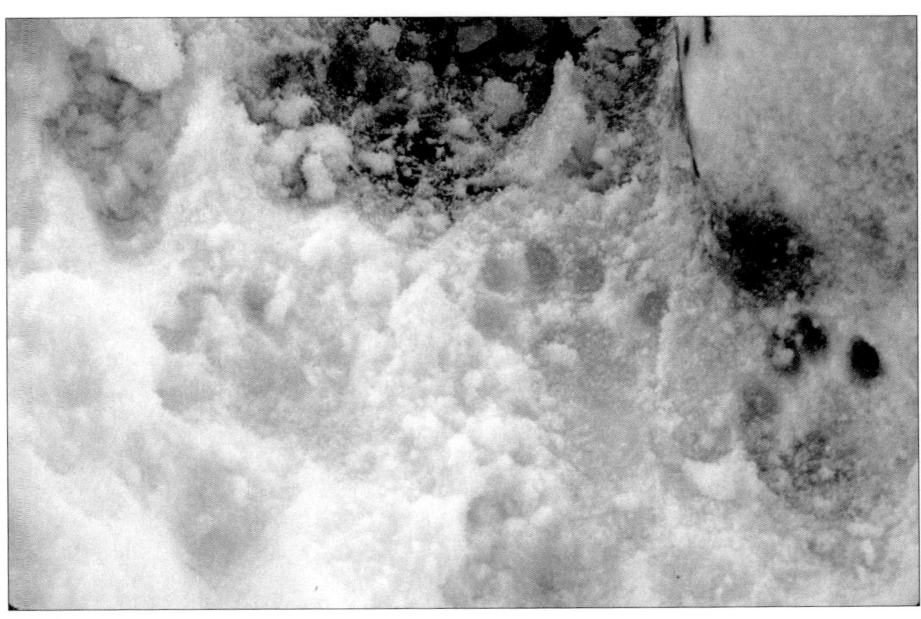

The otter tracks lead to a slide that empties into a hole in this ice-covered waterway. The beaver tracks lead back to a small segment of open water in their pond.

Looking out onto the nearby river, open waters are interspersed with patches of ice. In fact, with increased melting, ice regularly crashes into the river's flow sounding like a breaching whale splashing back against the ocean's surface.

I reflect on the new world awaiting all montane species here in the Colorado Rockies, since this year's snow pack was 150% of normal. Just think of the work that lies ahead for our beaver family with repairs to make during the maintenance of their dams.

Not to worry about their work force, though. Beaver families achieve a high degree of excellence and efficiency through mutual cooperation and coordination of activities with a prescribed division of labor. As we already discovered on our journey, all family members (except the kits) cooperate in building and maintaining the family's lodges and dams and stocking the winter food cache. Yearlings help adults raise the kits by provisioning them with food, babysitting, and protection. Also, they assist adults in cleaning the floor of the lodge by pushing soiled litter material into the water hole. Then new bedding and litter material is brought into the lodge and even created by shredding wood from branches.

During all seasons of the year, family members assist each other with mutual grooming behaviors to keep all areas of their fur water-repellent. Then particularly during the long winter months inside the close quarters of the lodge, each individual performs behaviors that ensure all members of the family are kept healthy and warm.

All their work as a family unit is very impressive!

Soon the yearlings will be moving into a different world. When they reach their second birthday, they will begin a life independent of their parents and other siblings. However, the former kits will themselves become yearlings, reaching a new level in their lives. We still have a month or so before all this occurs.

In the meantime, the melting ice on the beaver pond has brought the family forth from their winter isolation.

"Hi," I sing out to my special beaver. "Although we still have a lot of snow, you can at least emerge from the confinement of your lodge under the thinning layer of ice."

"Yes," responds Beaver. "It feels good to the entire family to stretch our muscles and breathe fresh air."

"I am so happy to see that you have all survived our difficult winter season," I speak with a voice of jubilation.

"I agree with your words," articulates Beaver. "Since my yearling youngsters will soon reach breeding age, their dispersal will be imminent. During their second year of life, they developed and refined the necessary building behaviors that will help them when they find their own 'homestead.' Also, by helping care for their younger siblings, they prepared themselves for parenthood. They have learned very well."

"I will ask Mother Nature to guide them on their journey," I express contemplatively.

"Thanks," voices Beaver.

Coincidental with beaver dispersal is another important aspect of beaver behavior: territorial marking. In our area, this occurs primarily from mid-spring into at least early summer. Marking behavior lets dispersing beavers know what properties are already taken. We will learn some of the details of dispersal and marking as our journey continues.

* * *

Springtime in the Rockies! What a dynamic time!

Each day presents a new surprise and a new challenge. As the snows and ice begin to liquefy, water seems to be everywhere. I look out onto the flow of the river where both my visual and aural senses are treated to this spectacular event, as the waters overflow their banks to flood the adjacent woodland.

However, progressing through April and into May does not mean the end of cold ambient temperatures and snow. Old Man Winter just doesn't want his season to end.

"Hey," I yell out as snow belts my being. "That's enough. Migrant birds will soon be arriving; they don't need more of this white stuff. Plus, my beavers have a lot of work to do to maintain their pond levels. So, maybe you could let up for now. It should be obvious that when you finally melt, we'll have more than sufficient life-giving water this year. Just think, you will return again in the fall."

The answer comes in the form of soft flakes of white, albeit falling rather sparsely.

Returning to the beaver pond, water is beginning to flow over their large dam as the river continues its rise. The family comes to do some work.

"I asked Old Man Winter to refrain from snowing, but he won't relent," I tell my special beaver.

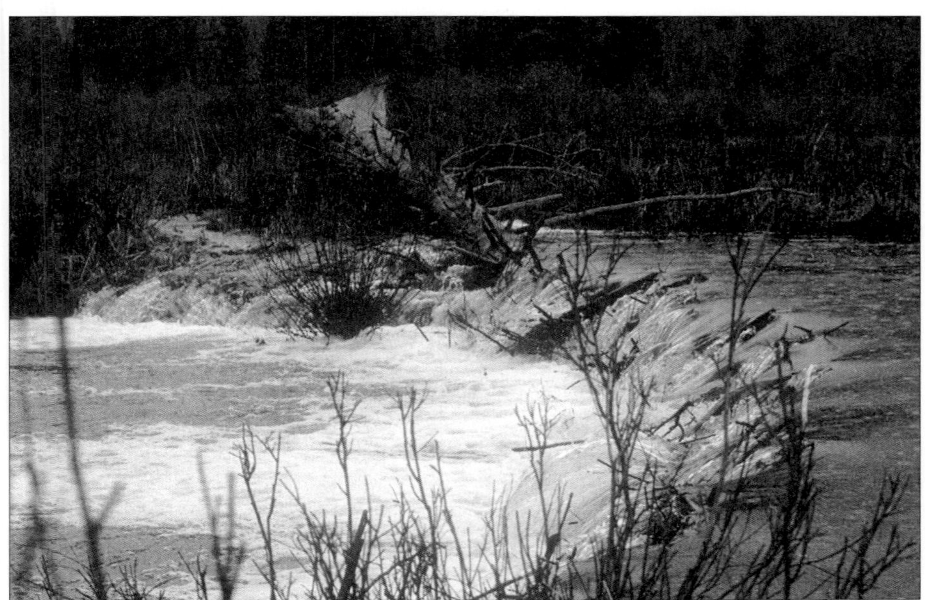

"Thanks for trying, but we are used to varying conditions in our home environment," Beaver responds. "During spring melt, if our major lodge should get flooded, we have another nearby lodge to use until Nature's great waters begin to recede. In the meantime, we will try to work on our dams. Look, our former kits are now helping out. Building is an important part of their learning process."

"You have a lot of work to do, so I'll stay out of your way," I express. Now cold and wet, I sit cuddled by the roots of a large tree to observe the beavers' behaviors.

A break in the clouds reveals the distant snow-covered mountains, their peaks peering over their pine-endowed slopes. Nearby, willows and grasses are emerging from the snowy landscape, still dressed in their grays and browns of winter. Ah, but there are hints of spring, as some green shoots begin peeking through the snow-covered ground, while small buds dot the bonnets of willows. Spring is really trying to overtake winter!

As the sun peeks in and out from behind the clouds, I gaze out onto the large beaver pond and think about the many advantages this one species' work provides. We have already discovered many benefits derived from beavers' activities, but there are even more.

Their two major ecological roles in the web of life are as ecosystem engineers and as keystone species in wetland and riparian ecosystems.

Let us summarize what this means. First, wetlands have been referred to as the "kidneys of the landscape" because they cleanse water of pollutants. Our beaver engineers help create wetlands. However, when humans drain this habitat and kill off the beavers, they are left with what they deserve: worsened floods, erosion, and a degraded ecosystem from lowered water tables. The effects of climate change — itself largely anthropogenic— will only amplify these human-created problems.

Further benefits of beaver dams on rivers and streams include trapping sediment, attenuating seasonal water-table fluctuations, stabilizing banks, controlling run-off, elevating water tables, improving growth of willows and other riparian plants, and creating habitat for many other faunal species, a few of which we already discovered on our journey. By trapping sediment, dams also trap phosphorus and nitrogen, which improves downstream water quality.

In locations where beavers move from one area to develop another, the succession of habitats, themselves, provide benefits to a variety of species, until once again the circle of life returns. However, here where I'm located, and in many known areas of the country where beavers are allowed to continue making their life-giving contributions, there is no observable reason for them to leave.

A lot of scientific detail is involved in each of these life-giving enhancements from our beavers, but this gives you, the reader, a general outline of this species' role and contribution to the balance of Nature.

* * *

Mud season has arrived, along with my progression from snowshoes to water boots. Now on the mud, I slip and slide onto my rear, which I must abide. Not only are there slippery, muddy soils to deal with, but also unsuspecting strong undercurrents that hide in the ebbs of the now-swiftly-flowing river.

Sometimes while trying to avoid mud and thick willow entrapments, I walk into the shallows of these waterways.

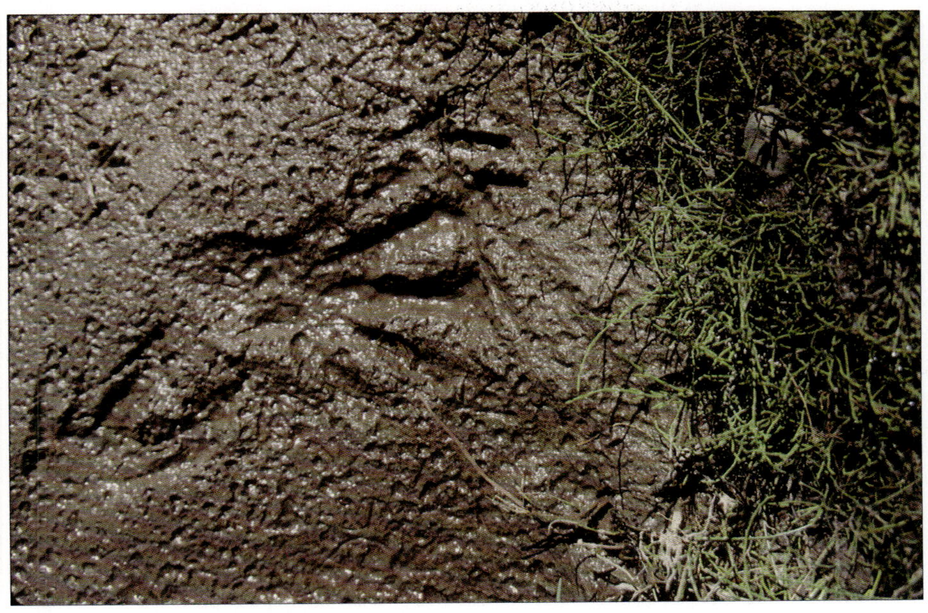

Not a good idea. The river's swift current tries to pull my legs out from under my small body towards its depths; I struggle to maintain my footing. I am not complaining; this is just how it is.

Actually, all this water provides a safe haven for our dispersing two-year-olds. They have begun venturing out on short excursions along the river and tributary creeks to determine if there are any nearby vacancies in good beaver habitat. Because they are challenged by predatory confrontations, having all this water around them provides them with a safety zone for retreat, if necessary. The water also keeps them cool during their travels. In addition, abundant waterways mean abundant banks to create holes for hiding and resting.

How far young beavers must travel depends on available habitat, which also depends on population density. This can mean less than a mile to several miles. In locations of degraded habitat, females have been found to travel farther from their natal family than males. It is hypothesized that they need more and better quality food for successful breeding. Therefore, dependent on habitat conditions, they may have to venture farther to find it. Males may not be as dependent on quantity and quality of food.

Whatever the case, during their dispersals, individuals of both sexes may find their future mates. Then they can set up "housekeeping" together. A collateral benefit of this dispersal disparity is an increase in variation of the species' gene pool.

Realistically, the picture is not all rosy. From extensive research in various areas of our country, most dispersers die before reaching their final destination. However, with so much good habitat in my area of the Rockies, I am encouraged that survival for our young beavers will be assured.

Dispersing two-year-olds may take weeks or even months to find their own territories. In addition to the benefits afforded them by abundant water, they may receive additional help along the way from their relatives. These young beavers may visit their kin for a few weeks before finally moving on to their yet unknown destination.

How do these beavers know that they are related? Secretion from their anal scent gland provides this information.

Beavers have an extremely sensitive nose so chemical communication is very important to their species. They have two different types of scent structures: one pair of anal glands and one pair of castor sacs. From each is deposited secretions that have various meanings within the species. Both sets of glands are located in two cavities between their pelvis and the base of their tail opening into the urogenital pouch.

There may be 100 or more different chemical compounds contained in the anal scent gland secretion. One function of this secretion is identification. Research results have shown that information about kinship is coded within this gland's secretion. An individual distinguishes its relatives from non-relatives by comparing its own scent with that from another encountered beaver. A beaver also distinguishes an individual's sex from the secretion by testing its color and consistency. In both cases, the secretion must be rubbed onto a substrate. Finally, it is also thought that the oily substance from this gland can be rubbed into their fur to assist them with water-proofing it.

Now we come to the more familiar gland — from which they get their genus name — the castor scent gland, which contains the substance castoreum. The secretion from this gland is also composed of over 100 different chemical compounds. Unlike the anal scent gland, the beaver does not have to physically contact the substrate to deposit the secretion. We'll call upon our special beaver to help us out with its behavioral purpose.

"Hi," I sing out to beaver. "Why are you making mud pies?"

"I don't understand the term 'mud pie'," responds Beaver in a quizzical manner. "What I am doing is creating scent mounds. First, I bring mud up from the bottom of my pond, then deposit it on land near the bank. In particular, I place the mud mounds, along with some aquatic plants and grasses, near my dams, lodges, canals, and trails to feeding locations. Once built, I apply and reapply castoreum to them. This lets any dispersing or meandering beavers know that this property is already taken."

"Also, by placing your chemical secretions on these moist, raised mounds, you both intensify the odor, due to their moisture content, and keep them from being flooded from water-level fluctuations, due to their raised level," I chime into the discussion.

"Well, I just do what comes naturally," Beaver continues. "As to my territorial marking, I seem to be the primary worker, although I do receive some help, particularly from my mate. Most of this work must be conducted during late spring and early summer when dispersal of two-year-olds is in full swing."

"In trying to understand the complexity of your species, I have found information that I would like to share with you," I state. "Doing what comes naturally is great for you, but we scientists try to understand the details. Did you know that many of the chemical compounds found in your castoreum actually come from the foods you eat?"

"That's interesting," responds Beaver. "Maybe that is why we are called a 'choosy generalist herbivore.' "

"Touché," I continue. "Oh, it doesn't end there. Briefly, many plants try to defend themselves from you and other herbivores, by developing strong chemical defenses. However, your species has adapted to this defense by physiologically using these plants' adverse chemicals in your own castoreum, thus recycling them as signals for territorial marking."

"I guess your words 'evolution' and 'adaptability' continue to be prominent factors in our lives," speaks Beaver.

"Yes, that is so true," I contemplate. "Before you continue with your duties and feeding behavior, I have one more piece of information to share with you. The word 'recycling' brought to mind an interesting behavior I once witnessed. Before you and I connected, I followed a different family up here in the Colorado Rockies and was privileged to observe them removing branches from one of their old dams and moving them downstream to the construction site of a new one over a period of a few nights. So, in addition to the multitude of benefits your species provides to the web of life, you are also eco-sensitive!"

"I will have to remember that behavior for any future building projects," speaks Beaver. "Thanks for letting me know. Now I must continue with my intense workload."

* * *

The next day, I embrace this latter part of spring by venturing to other locations along the Colorado River. Looking out onto these crystal clear waters tinged with blue as they flow with unrelenting force to reach their goal, I think about the mysteries that await them along their path of travel. I think about the life-giving resource they will provide to both the flora and the fauna they encounter on their journey.

Gazing across the landscape, I watch as Mother Nature busily spreads her display of floral hues across the blooms of life, looking like the assortment of colors from the palette of artist Vincent Van Gogh.

"*You have a wonderful array of flowers to brighten your warm season bonnets*," I sing out to Mother Nature.

A gentle wisp of wind kisses my cheek in response, followed by the harmony of sounds from her avian chorus.

Now entering their breeding season, species of birds sing their songs of love to attract mates and begin a new generation of flying beauties. The tiny hummingbird flitters above my head perhaps to pull some golden hairs for its nest. I love watching their flights of fancy. This one small wonder brings forth a feeling of peace and harmony within the Natural World.

Some species of fauna have already born their young and are beginning to acquaint them with the environment they will call home. Mother Nature must feel proud of the

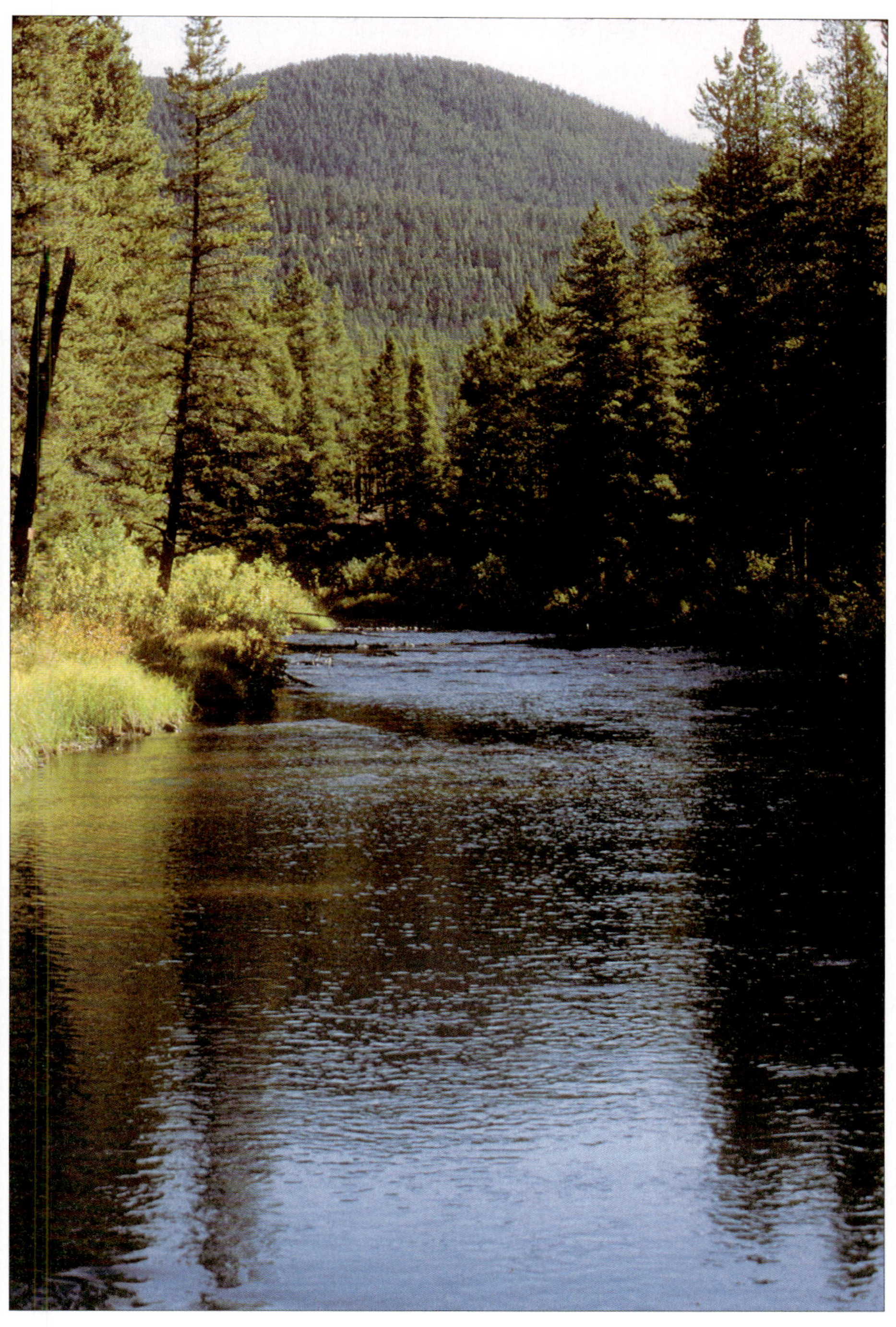

many species being introduced for her approval as she ensures their debuts are draped in sunshine. Not to be outdone by fauna and pretty flora, foliates of deciduous trees, shrubs, and other plants are also emerging for their acceptance in the natural scheme of life.

At every turn in this scene from Nature is found the timeless beauty of her spring awakening.

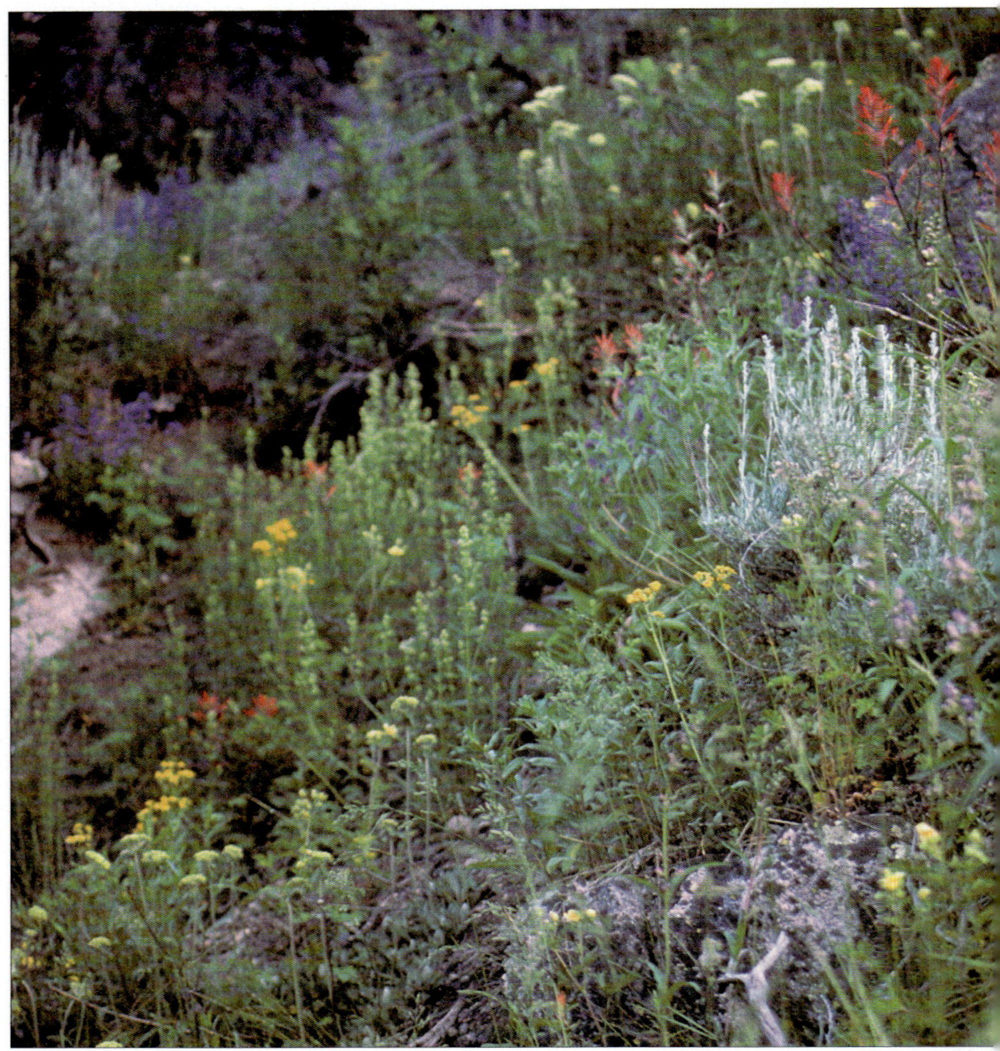

Returning to the beaver pond, I look out onto its quiet, shimmering waters as they reflect newly-foliated willows waving in the gentle breeze and bonnies of color shyly peeking their tiny faces through shapely petals. The snow-capped mountain peaks proudly lend their own beauty to this mirror image, while above, the sky's intense blue is broken only briefly by billowy shapes of white cotton candy and Nature's winged creations.

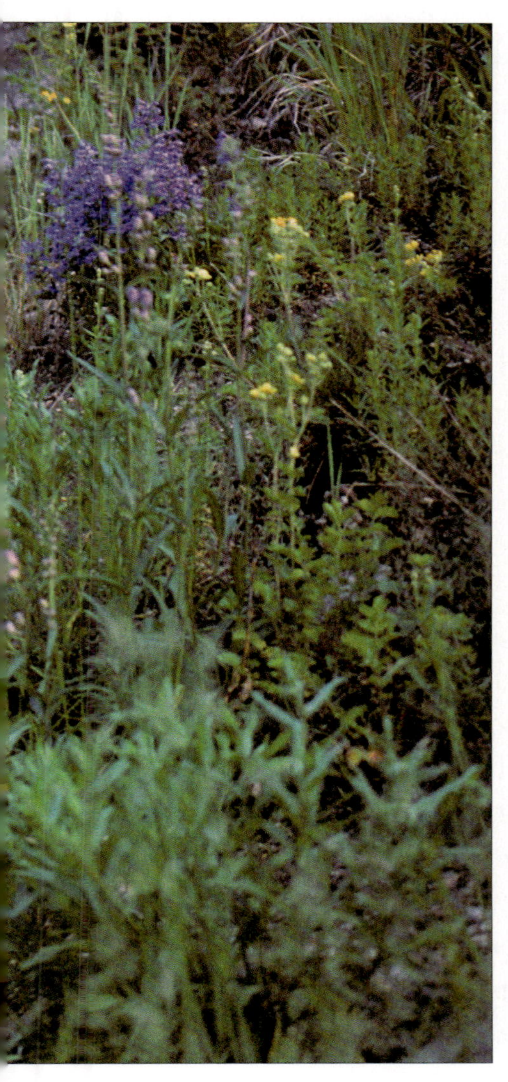

Here come the swallows to dip into these waters for flying insects. The ducks land to create a distraction in these still waters and search for their own nourishment. Up pops a fish for its meal before becoming one itself, since I just spotted a river otter. There comes a kingfisher to also partake in the bounty. This pond is a good place for fish to spawn and to grow up, before becoming a major resource to various species.

I search the sandy shores along these waterways where clues to the movements of some of Nature's children are revealed as impressions in the mud. There are so many species that benefit from this beaver pond. No matter their size, they sing out a big "thank you" to the beaver . . . and so do I.

My time in this beautiful environment will soon be drawing to a close. My special beaver told me that a new generation of kits has entered our world. I have now completed a full seasonal cycle in the life of this one family.

"Kheeeeer" resounds throughout my world.

"Yes," I respond to the red-tailed hawk. *"I have learned and felt beyond what I observed in Nature, just as you told me in your original message."*

* * *

I return to the field of sunflowers whence our story first began. Once again, they invite me to join them in their dance of summer. Swirling amongst their cheery faces, I finally fall down upon the earth from exhaustion. Although soon my final journey up into the Rockies must begin, first I will choose one of these flowers to take with me.

"Who amongst you would like to become the chosen one as a gift to my special beaver family?" I sing out to the golden beauties.

Looking up from my reclined position, I gaze into the face of one large bloom that stands out above the rest. I feel it beckoning me to select it as the chosen one. This particular flower reminds me of Vincent Van Gogh's 1887 painting of a huge sunflower growing at the edge of a Parisian garden, its large green leaves positioned like sentinels encircling a tall stem crowned with an enormous bloom, just like the one I now pick.

Returning to the Rocky Mountains, I grasp onto the sunflower and begin a slow walk to my beaver's pond.

Internally, I feel both happy and sad. Happy reflecting that my special beaver taught me so much about his species' natural history and that we shared each other's information about *Castor canadensis* (mine from personal observation and the scientific literature, and his from living it), all making me feel an integral part of his world. But I feel sad because, upon my departure, we may never see each other again.

"Hi," I sing out as my beaver emerges from the sparkling blue waters of his pond.

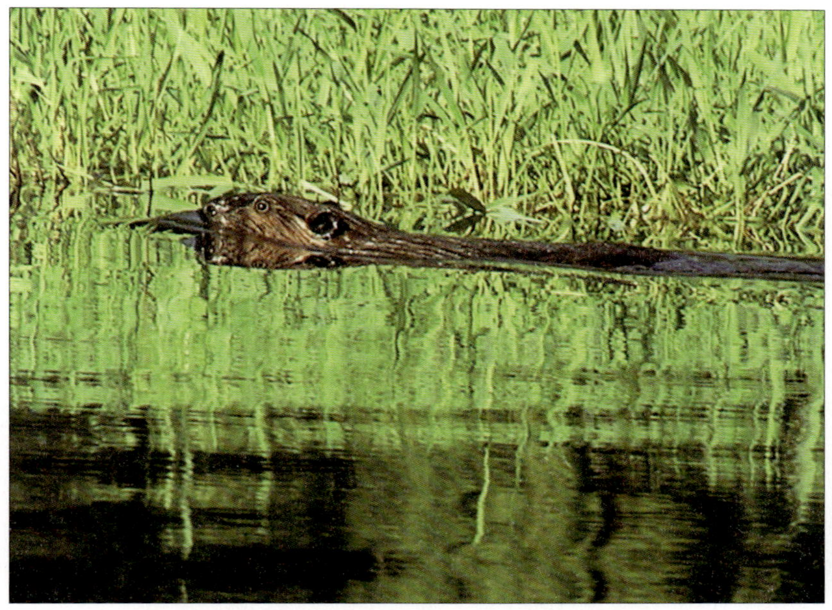

"Glad to see you again," responds Beaver.

Speaking in a tearful state, "I brought you a sunflower gift."

The beaver moves over to me, and we touch. Then he grasps the giant sunflower and walks back to his pond. Before entering the waters, he turns and speaks to me.

"Do not be sad," emerges from his voice. "You must now share the story of our species with the human world. Then, just maybe, people will become enlightened about beavers and try to do what they can to ensure our survival for eons in the future as we survived for eons in the past."

"I promise to do my best," I speak in a heartfelt manner to him. "Good luck to you and your family."

"Thanks, and same to you and yours," he speaks before entering the waters with the sunflower grasped in his jaw.

Although it's a bright sunny day, a layer of fog swoops down from the mountain tops and envelops his pond. Maybe this is a message from Mother Nature that she will continue *trying* to protect him and his species.

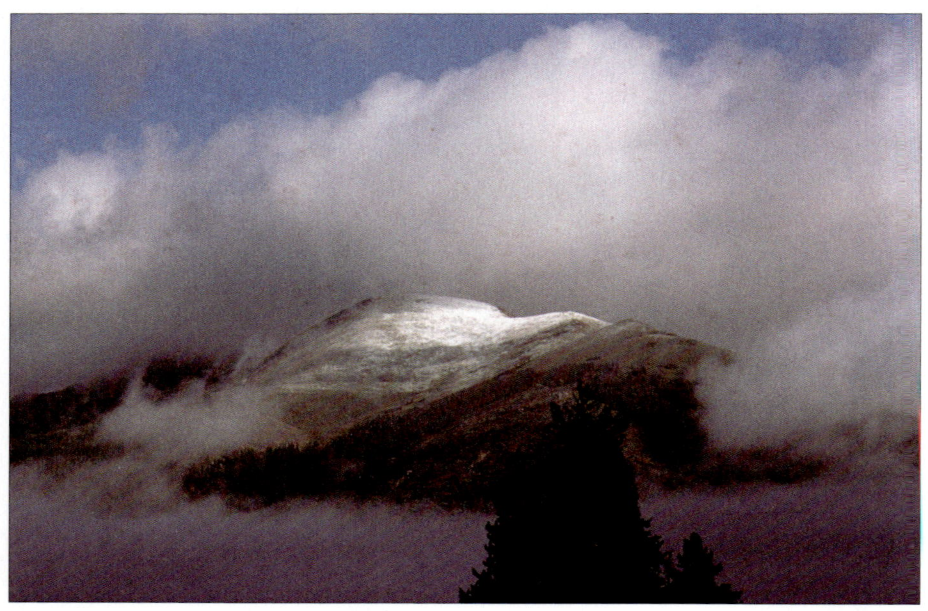

Then as I turn to walk away, a wisp of wind kisses my cheek. Smiling, I respond, "Thank you, I too will *try*."

"Kheeeeer" resounds from the red-tailed hawk.

References

Aleksiuk, M. 1968. Scent-mound communication, territorality and population regulation in beaver (*Castor canadensis*). *J. Mammal.* 49:759-762.

Aleksiuk, M., and I. McT. Cowan. 1969. Aspects of seasonal energy expenditure in the beaver (*Castor canadensis*) at the northern limit of its distribution. *Can. J. Zool.* 47:471-481.

Armstrong, D.M. 1975. *Rocky Mountain Mammals*. Rocky Mtn. Nature Assoc., Inc. CO.

Berg, J.K. 1999. *Final report of the river otter research project on the upper Colorado River basin in and adjacent to Rocky Mountain National Park, Co*. National Park Service.

Berg, J.K. 2000. North American river otter diet. *River Otter Journal*. Vol. IX No. 2:4-5.

Brady, C.A., and G. E. Svendsen. 1981. Social behavior in a family of beaver, *Castor canadensis*. *Biol. Behav.* 6:99-114.

Breck, S.W., K.R. Wilson, and D.A. Andersen. 2001. The demographic response of bank-dwelling beavers to flow regulation on the Green River. *Can. J. Zool.* 79:1416-1431.

Buech, R.R., D.J. Rugg, and N.L. Miller. 1989. Temperature in beaver lodges and bank dens in a near-boreal environment. *Can. J. Zool.* 67:1061-1066.

Buech, R.R. 1995. Sex differences in behavior of beavers living in near-boreal lake habitat. *Can. J. Zool.* 73:2133-2143.

Busher, P.E., and S. H. Jenkins. 1985. Behavioral patterns of a beaver family in California. *Biol. Behav.* 10:41-54.

Cole, R.W. 1970. Pharyngeal and lingual adaptations of the beaver. *J. Mammal.* 51:424-425.

Dyck, A.P., and R. A. MacArthur. 1993. Seasonal variation in the microclimate and gas composition of beaver lodges in a boreal environment. *J. Mammal.* 74:180-188.

Elbroch, M. 2003. *Mammal Tracks & Sign: A guide to North American species*. Stackpole Books. PA.

Feldhamer, G.A., B.C. Thompson, and J.A. Chapman, Eds. 2003. *Wild mammals of North America: biology, management, and conservation*. Johns Hopkins U. Press: Baltimore, MD.

Hodgdon, H.E., and J.S. Larson. 1973. Some sexual differences in behavior within a colony of marked beavers (*Castor canadensis). Anim. Behav.* 21:147-152.

Jenkins, S.H., and P.E. Busher. 1979. *Castor canadensis. Mammal. Species*, 120:1-8.

Lancia, R.A., W.E. Dodge, and J.S. Larson. 1982. Winter activity patterns of two radio-marked beaver colonies. *J. Mammal.* 63:598-606.

Lariviere, S. and Walton, L.R. 1998. *Lontra canadensis. Mammal. Species*, 587:1-8.

MacArthur, R.A., and A.P. Dyck. 1990. Aquatic thermoregulation of captive and free-ranging beavers (*Castor canadensis*). *Can. J. Zool.* 68:2409-2416.

Muller-Schwarze, D., and Heckman, S. 1980. The social role of scent marking in beaver (*Castor canadensis*). *J. Chem. Ecol.* 6:81-95.

Muller-Schwarze, D., and P.W. Houlihan. 1991. Pheromonal activity of single castoreum constituents in beaver, *Castor canadensis. J. Chem. Ecol.* 17:715-734.

Muller-Schwarze, D., and B. A. Schulte. 1999. Behavioral and ecological characteristics of a "climax" population of beaver (*Castor canadensis)*. Pp. 161-177 in *Beaver protection, management and utilization in Europe and North America* (P. E. Busher and R. M. Dzieciolowski, eds.) Kluwer Academic/Plenum Publishers, New York.

Muller-Schwarze, D., and L. Sun. 2003. *The Beaver: Natural History of a Wetlands Engineer.* Cornell U. Press. Ithaca, N.Y.

Neff, D.J. 1957. Seventy year history of a Colorado beaver colony. *J. Mammal.* 40:381-387.

Novak, M. 1987. Beaver. Pp. 283-312 in *Wildlife furbearer management and conservation in North America* (M. Novak, J.A. Baker, M.E. Obbard, and B. Malloch, eds.) Ashton-potter Limited, Ontario Ministry of Natural Resources, Concord, Ontario, Canada.

Novakowski, N.S. 1967. The winter bioenergetics of a beaver population in northern latitudes. *Can. J. Zool.* 45:1107-1118.

Nowak, R.M. 1999. *Walker's Mammals of the World.* Sixth Edition, Vols. I and II. Johns Hopkins U. Press. Baltimore, MD.

Nowak, R.M. 2005. *Walker's Carnivores of the World.* Johns Hopkins U. Press. Baltimore, MD.

Patenaude, F. 1984. The ontogeny of behavior of free-living beavers (*Castor canadensis*). *Z. Tierpsychol.* 66:33-44.

Patenaude, F. and J. Bovet. 1984. Self-grooming and social grooming in the North American beaver, *Castor canadensis. Can. J. Zool.* 62:1872-1878.

Petersen, D. 1988. *Among the elk.* Flagstaff, AZ. Northland Pub.

Polis, G.A., M.E. Power, and G.R. Huxel. 2004. *Food webs at the landscape level.* University of Chicago Press: Chicago, IL.

Rezendes, P. 1993. *Tracking & the Art of Seeing: How to Read Animal Tracks & Sign.* Camden House Publishing, Vermont.

Rosell, F. and B.A. Schulte. 2004. Sexual dimorphism in the development of scent structures for the obligate monogamous Eurasian beaver *(Castor fiber*). *J. Mammal.* 85:1138-1144.

Samuels, J.X. and J. Zancanella. 2011. An early Hemphillian occurrence of Castor (Castoridae) from the Rattlesnake Formation of OR. *J. Paleontology*: 85:930-935.

Savage, R.J.G., and M.R. Long. *Mammal Evolution: an illustrated guide. Facts on File Publication*, Oxford, England.

Schulte, B.A., D. Muller-Schwarze, and L. Sun. 1995. Using anal gland secretion to determine sex in beaver. *J. Wildlife Mgnt.* 59:614-618.

Schulte, B.A., D. Muller-Schwarze, R. Tang, and F. Webster. 1995. Bioactivity of beaver castoreum constituents using principal components analysis. *J. Chem. Ecol.* 21:941-957.

Schulte, B.A. 1998. Scent marking and responses to male castor fluid by beavers. *J. Mammal.* 79:191-203.

Singer, F.J., L.C. Zeigenfull, R.G. Cates, and D.T. Barnett. 1998. Elk, multiple factors, and persistence of willows in National Parks. *Wildlife Soc. Bulletin.* 26:419-428.

Stephenson, A.B. 1969. Temperatures within a beaver lodge in winter. *J. Mammal.* 50:134-136.

Stein, B.R. 1989. Bone density and adaptation in semi-aquatic mammals. *J. Mammal.* 70:467-476.

Sun, L., D. Muller-Schwarze, and B.A. Schulte. 2000. Dispersal pattern and effective population size of the beaver. *Can. J. Zool.* 78:393-398.

Sun, L., and D. Muller-Schwarze. 1997. Sibling recognition in the beaver: a field test for phenotype-matching. *Animal Beh.* 54:492-502.

Svendsen, G.E. 1980. Patterns of scent-mounding in a population of beaver (*Castor canadensis*). *J. Chem. Ecol.* 6:133-148.

Svendsen, G.E. 1989. Pair formation, duration of pair-bonds, and mate replacement in a population of beavers (*Castor canadensis*). *Can. J. Zool.* 67:336-340.

Tevis, L. 1950. Summer behavior of a family of beavers in New York State. *J. Mammal.* 31:40-65.

Townsend, J.E. 1953. Beaver ecology in western Montana with special reference to movements. *J. Mammal.* 34:459-479.

Van Gelder, F.G. 1982. *Mammals of the National Parks*. Johns Hopkins U. Press. Baltimore.

Welsh, R.G. and D. Muller-Schwarze. 1989. Experimental habitat scenting inhibits colonization by beaver, *Castor canadensis*. *J. Chem. Ecol.*, 15:887-893.

Williams, C.L., S.W. Breck, and B.W. Baker. 2004. Genetic methods improve accuracy of gender determination in beavers. *J. Mammal.* 85:1145-1148.

Other References

Vincent Van Gogh: Information from letters to his brother, Theo:

Roskill, Mark. 1963. *The letters of Vincent Van Gogh*. Atheneum Edition, New York. (This edition contains a selection of Van Gogh's letters that appear in a three-volume edition published by Constable in 1927/1929.).

Other Van Gogh-related references that particularly inspired me:

Bonafaux, Pascal. 1992. *Van Gogh: The Passionate Eye*. Thames & Hudson/New Horizons, London & New York.

Skea, Ralph. 2011. *Vincent's Gardens: Paintings and Drawings by Van Gogh*. Thames & Hudson, New York.

Royal Academy of Arts. 2010. *The Real Van Gogh: The artist and his letters*. From an exhibition in 2010 by the Royal Academy of Arts, London.

Native American Legends:

"Beaver and Porcupine." Told by Brown Bear Mallot, a spiritual leader from the Tlingit Tribe in Alaska. In Lake-Thorn, Bobby. 1997. *Spirits of the Earth: a guide to Native American Nature symbols, stories, and ceremonies*. Plume/Penguin Books, New York.

"The Bluebird and Coyote." Told by Frank Russel in 1908, of the Pima Tribe in Arizona. In Erdoes, Richard and Alfonso Ortis. 1984. *American Indian Myths and Legends*. Pantheon Books, Div. of Random House, New York.

Willow Creek and Fern Ridge

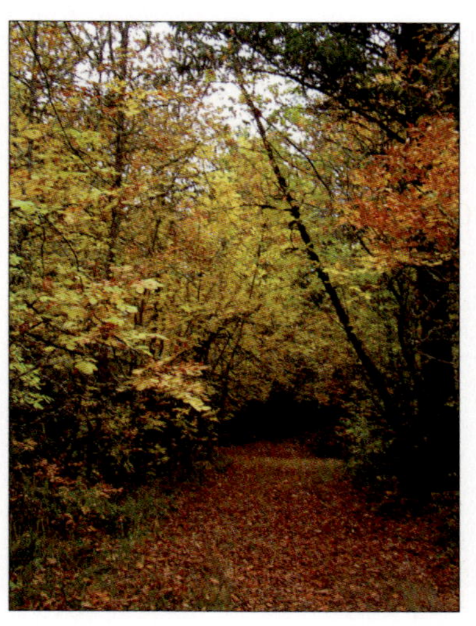

PREFACE TO
Willow Creek and Fern Ridge

The next two natural history journeys are taken from accounts of my activities conducted in special places close to home: Eugene, Oregon. Willow Creek Preserve is the site of a four-year mammal survey I conducted from 2006 through 2009; Fern Ridge Lake and one of its tributaries, Coyote Creek, are favorites for frequent spring, summer, and fall canoeing to observe and listen to breeding birds, amphibians, fish, and the occasional mammal along protected wetlands on its eastern shore. The following demographics will help establish the geographic context of the stories.

Eugene, Oregon, is situated in the southern Willamette River Valley, at an elevation of 430 feet above sea level. It is 112 miles directly south of Portland, roughly centered between the Cascade Mountains to the east and the Coastal Range to the west. Along the eastern and southern flanks of the city is the Eugene Ridgeline, a series of adjacent hills ranging up to 2,200 feet in elevation. The Ridgeline's west- and north-facing watershed flows into Amazon Creek, a perennial waterway that meanders from southeast to northwest through Eugene, ultimately emptying into Fern

Ridge Lake, a reservoir along the Long Tom River, a tributary of the Willamette — itself, a tributary of the Columbia.

To the west of Eugene, along and on either side of Amazon Creek, is a 3,000-acre protected area known as the West Eugene Wetlands. Different portions of the Wetlands are owned and/or managed by a partnership of multi-level government agencies and The Nature Conservancy. Anchoring the Wetlands on the west is Fern Ridge Lake, an Army Corps of Engineers project encompassing 12,000 total acres, of which the lake, when full, occupies 9,000 surface acres. During the wet winter and spring seasons, the lake is one of several in the Willamette watershed used for flood control. Runoff from Ridgeline and Coast Range waterways ultimately fills the lake by the end of spring. During the dry summer, the lake is slowly drained for irrigation; it is finally emptied at the end of October in anticipation of a new cycle of winter rains. While portions of the lake are popular for recreation, much of the shoreline and surrounding wetlands are protected for the benefit of wildlife.

Willow Creek Preserve is a 508-acre site, owned by The Nature Conservancy, situated in the southern extent of the 3,000-acre watershed of the West Eugene Wetlands. It is the richest remnant of native wet prairie in Oregon's southern Willamette Valley. The preserve's raison d'etre is protection of the endangered Fender's Blue Butterfly (*Icaricia icarioides fenderi*) and its sole host plant, the threatened Kindaid's Lupine (*Lupinus sulphureus Kincaidi*). The Preserve includes wet prairie, upland prairie, Oregon white oak-California black oak woodland, Oregon ash forested wetland, riparian wetland, and forest communities.

Thus, in addition to the butterfly, it provides a diverse array of habitats for mammalian species as well. There are over 200 native plants plus many different species of fauna.

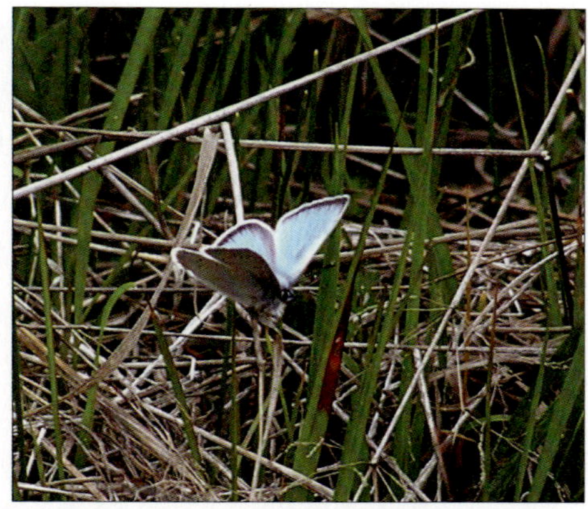

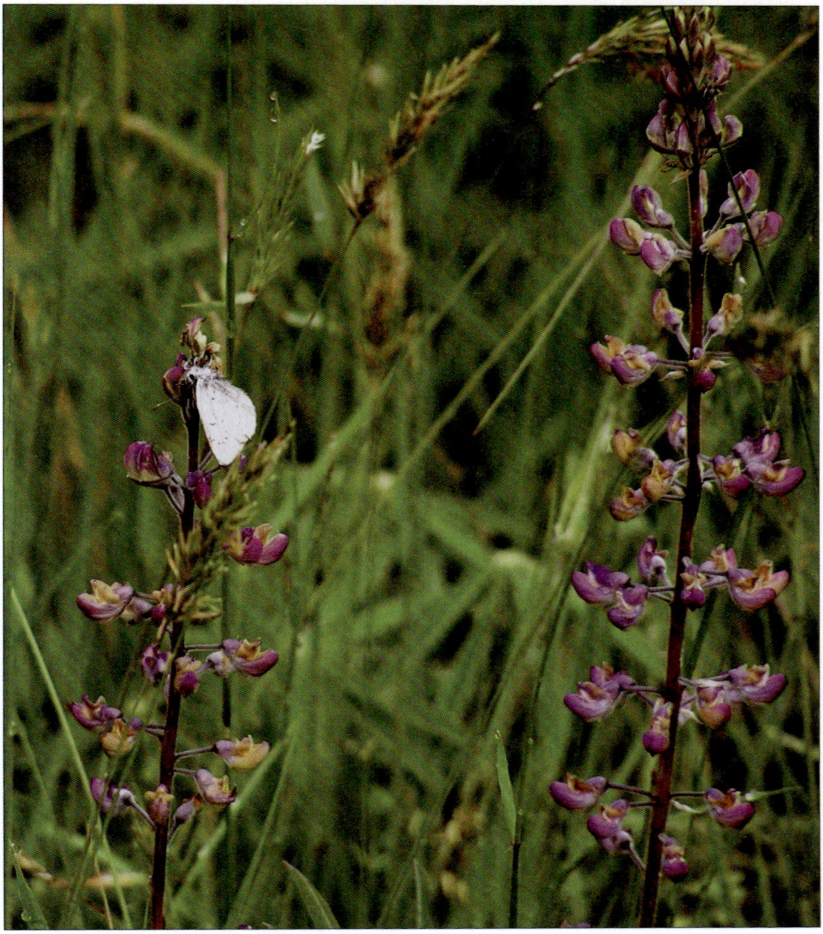

Key features of the Preserve are the seasonal East and West forks of Willow Creek, which drain two Ridgeline hillsides to the south and eventually converge to form a valuable component of the Wetland's Amazon Creek watershed. Runoff is insufficient to feed the Willow Creek forks year-round; they ultimately dry out during the dry summer and early fall.

From 2006 through 2009, I conducted a four-year volunteer mammal survey at Willow Creek Preserve to provide baseline data for future mammalian research and to determine if there were any seasonal variations in speciation or the addition of new species during this period of time. During that same period — and continuing to the present — my husband and I venture on weekly canoe trips during May through September across Fern Ridge Lake to the wetlands on its eastern shore, a paradise for nesting birds and other fauna.

Enjoy the following accounts of natural history to discover some mysteries of Nature found in the web of life.

Willow Creek Preserve

During late June, the coolness of early morning fills the air as I wind my way along the path of discovery. *"What mysteries will you unfold for me today?"* I ask Mother Nature. The melodious music from a song sparrow reverberates throughout my being, while smaller winged creatures embrace me in their light-hearted games.

These diversions are common throughout all months of the year while I conduct a mammal survey on Willow Creek Preserve. Even though I rarely see mammals, other faunal and floral species entertain my senses and imagination.

Patches of water that spread out in sections of the two seasonal creeks provide drink for the many faunal inhabitants, as evidenced by signs they leave behind. Other parts of the waterway have filled in with grasses of shades of green, dotted with sprays of color for Mother Nature's bonnets. Many choices present themselves to adorn her long flowing hair while matching a flock befitting her late spring fling.

As I search for signs that my mammalian friends left behind during their nighttime ventures, I discover scat (droppings) from one of my unsung carnivores. This is always of interest because it not only tells me which species deposited it, but also where, approximately when, and what had been its meal. I photograph, measure, collect it for analysis, and document indicative information. Tracks

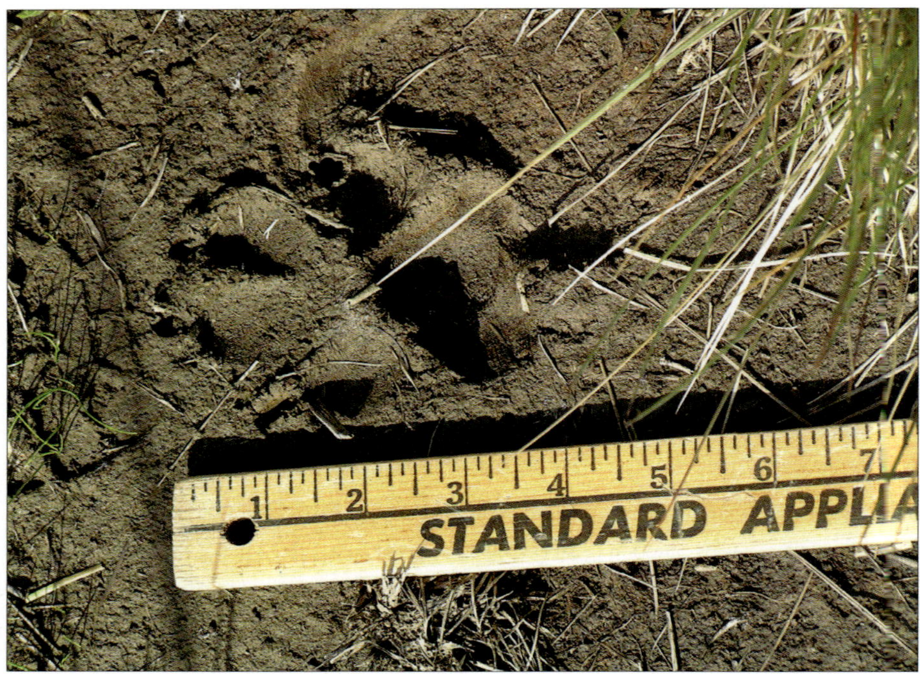

embedded in the wet earth nearby provide additional species confirmation.

Nature's mysteries abound in the Natural World, so if you like to play detective, what better job could there possibly be?

Educated evaluation reveals it was a coyote (*Canis latrans*) that left behind this "present." Analysis of its scat in my home laboratory reveals it was eating small members of the order *Rodentia* known as voles. Behaviorally, one can guess at what transpired. Coyotes hunt primarily by smell, especially when seeking small rodents such as mice and voles, but may use vision when hunting squirrels and rabbits. They normally use surprise rather than speed to capture their prey. However, in open areas and with larger prey, they can reach speeds of 40 mph.

During winter up in the Colorado Rocky Mountains, I once followed a Nature story, one of whose characters was a coyote. Tracks of the coyote had followed the tunneling behavior of a vole under the snow. The end of this story was apparent. Another time, I watched a coyote pouncing into a snow bank, a behavior their fox relatives also exhibit. In both cases, the prey was voles.

Recent evidence reveals a complex pattern occurring in mountainous regions where a good snow pack is prevalent throughout the colder seasons. Voles and other small rodents reside near the warmer earth underneath the snow — known as the "subnivean zone" — but their survival and reproduction are dependent on a complex interplay of biological systems influenced by ambient weather conditions. Their numbers can also affect the population of predators who rely on them as an important food source. With an increase in the number of small rodents, there will be a corresponding increase in their predator's numbers.

However, there is also a downside to large vole populations. Voles are herbivores (plant feeders) that have molars which grow throughout their lifetime. Therefore, since they feed on a variety of vegetation, including the more abrasive species, the more voles, the greater their live impact. This can be a real problem for endangered and threatened floral species. Their home ranges vary in size, dependent on species and habitat. For our two species in the Preserve, they are approximately 1,000 square yards for males and 400 square yards less for females. However, during periods of peak cycles, densities dramatically increase.

Much research has been conducted as to why there are vole cycles throughout all habitats in North America; the

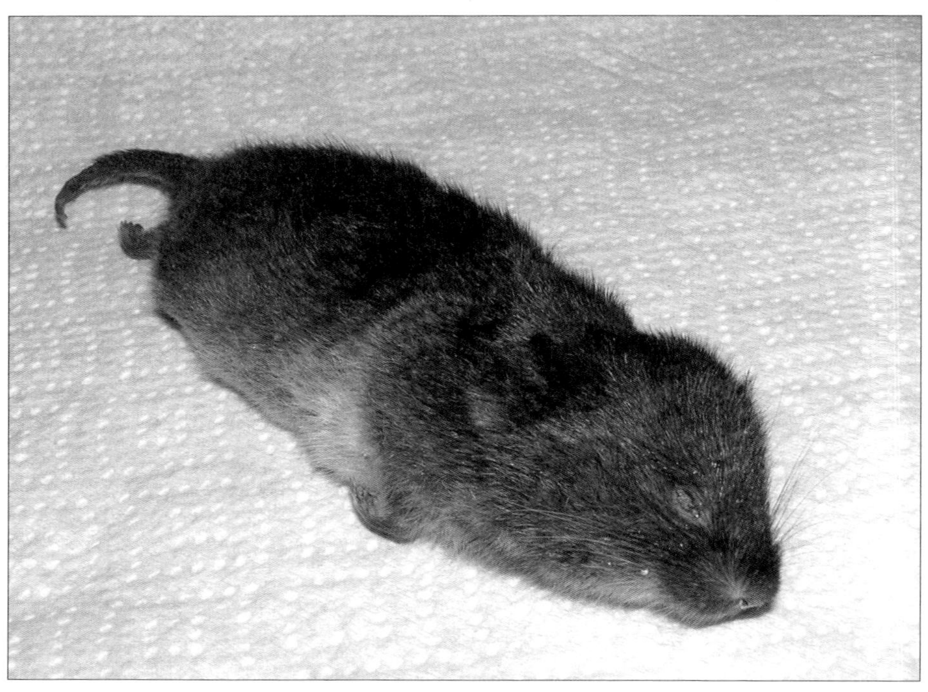

general consensus is that their fluctuations are based on intrinsic processes involving spacing behavior and extrinsic processes involving predation and food. In addition to population controls imposed by snow country, both mammalian carnivores and avian raptors predate on voles to keep their populations in check. Findings from my project in Willow Creek Preserve revealed the major prey items in carnivore scat throughout the year were the remains of our two species of voles: grey-tailed vole (*Microtus canicaudus*) and Townsend's vole (*Microtus townsendii*). Interestingly, Townsend's prefers wetter habitat than the grey-tailed. So, our predators have some habitat variety when seeking these species. The balance of Nature is at work.

Continuing on my path of discovery, tall grasses waving in gentle breezes give me a sense of freedom as I sway back and forth feeling their rhythm of life. As the warmth of the sun enters this scene, I sit upon the earth to watch Mother Nature's performance from her splendorous small- winged butterflies. Their colorful array highlights the flora upon which they are feeding from the nectars of life. I join them in their light-hearted games as my camera snaps their pictures for posterity. The endangered Fenders Blue butterfly and the flower that hosts it, Kincaid's Lupine, demonstrate the necessary connection to their life cycles. The Lupine is blooming in beauty for the butterfly to milk its kindness. Simplicity and complexity co-exist.

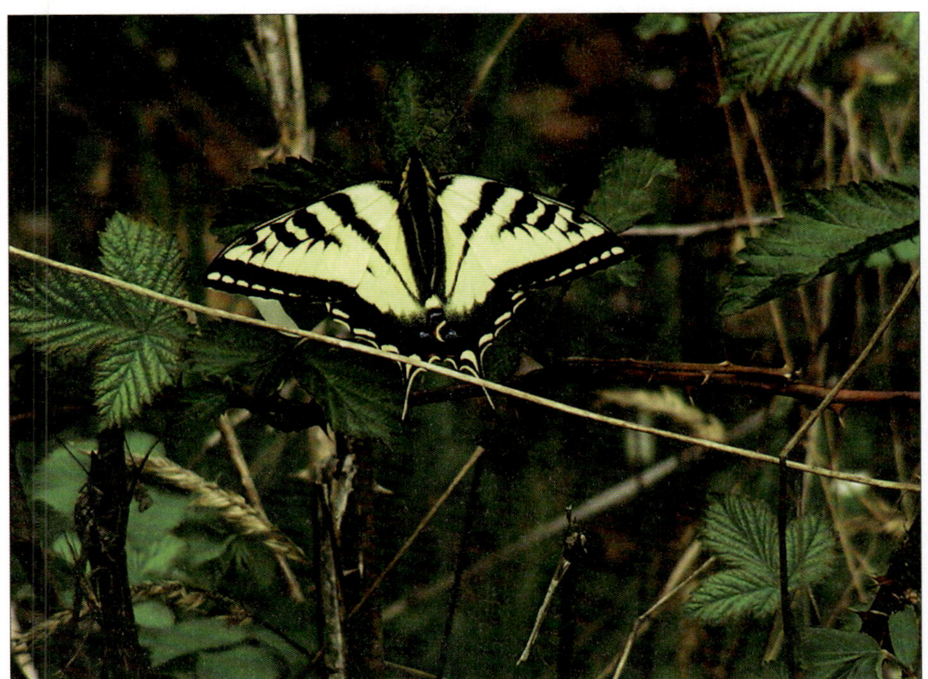

Into the woods, I follow a trail nicely carved out by black-tailed mule deer. Without their help, thickets of sharp needles would hinder my progress. Noiseless movement through the tops of tall trees pulls my visual senses skyward. A great-horned owl flies quietly overhead while spreading its mystery of magic, which trickles down like soft petals, surrounding me with its silent wisdom. Although usually a nighttime hunter, this bird may have just flown in response to an unknown stimulus. They are considered the most successful predator among the owls of North America, taking whatever prey presents itself.

The daytime correspondent to great-horned owls is the red-tailed hawk. They each occupy a temporal niche over the same territory — and they may even nest on the same tract of land. I often see red-tails soaring and gliding upon currents over the open meadows of the Preserve. To some Native American tribes, the red-tailed hawk is the messenger bird. I can attest to their being the bearer of guidance and enlightenment. During my research in the Colorado Rockies, a red-tail directed me to finding my special river otter, and another became a special messenger in my study of beavers.

Although my time here must end for now, there are more months and more adventures to come. Many more species will reveal some of their secrets.

* * *

With summer gaining a foothold, today I decide to enter the Preserve from a different and unfamiliar location. Walking in amongst densely packed grasses, I try to push my way through, since many are as tall as my five-foot height. Fortunately, I discover a nicely carved-out deer trail to follow. However, while intently looking for mammal signs, my movements take me away from the path. *"Gee, it must be somewhere,"* I speak to no one. Everywhere I turn, the habitat looks the same and nothing is familiar. *"Where is the path?"*

If I were a nonhuman animal, my strong senses would guide me out of this situation. Since that is not the case, panic begins to take over. I sit down upon the earth and ask Mother Nature to please extend me her guidance. Then a dark-colored butterfly appears and dances within the vegetation in front of me. *"Are you the sign I am looking for,"* I question the butterfly. *"Although you are much smaller than a red-tailed hawk, at least you found me."*

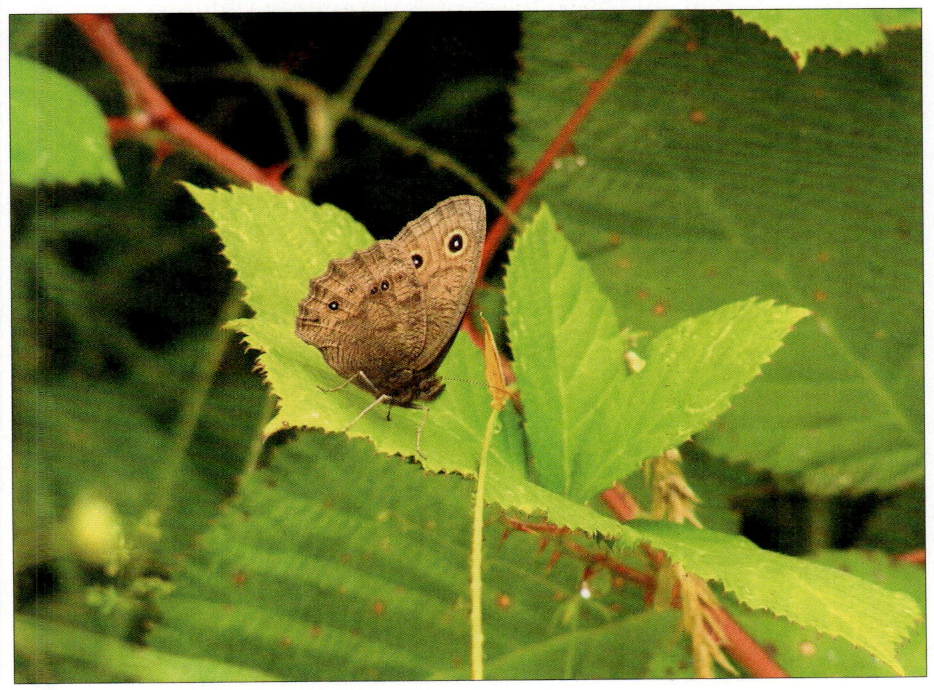

The small beauty of flight disappears, then reappears as if beckoning for me to follow. I stand and walk in the direction of its flight. As it dances amongst the grasses, I mimic its light-hearted behavior while continuing to follow its guidance. Eventually, I am returned to whence I began. The butterfly flutters in front of me, then disappears. "*Thank-you,*" I express to Mother Nature, *"for guidance from your small flittering beauty."*

* * *

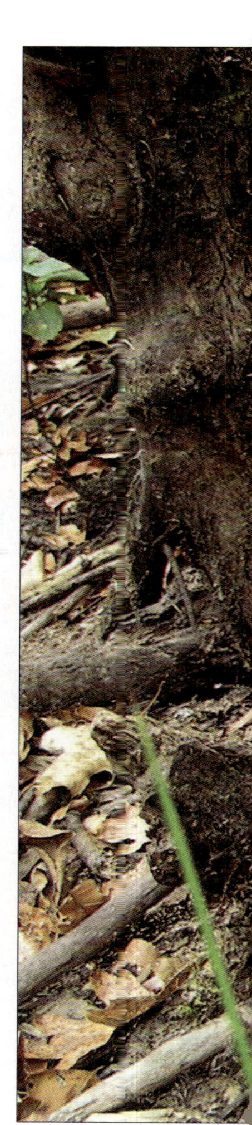

It is now August. The seasonal streams are primarily dry, except for a vegetative mat of green, which feels like a soft carpet as I walk across its cover. This drying-out period allows me to climb around locations that are flooded by waters flowing earlier in the year. Sweet mysteries of life are revealed as I move around in this little-known world. The heavily vegetated locations make it difficult to maneuver, but this detective desires to find out what clues are waiting to be discovered.

Lying within a bank and back under the roots of a large tree, a den for the semi-aquatic beaver (*Castor canadensis*) is unveiled. One opening has some sawdust in front of it. I haven't seen that before. My imagination pictures a beaver sawing wood to attain the correct size for its nearby dam. A different opening into a bank den unmasks some partially destroyed tracks embedded in the soil at its entrance. Enough of the tracks remain to tell me they were left by a beaver.

The semi-aquatic nutria (*Myocastor coypus*) is also known to reside here during the wet season, just like the beaver. Nutria are indigenous to the southern half of South America but were introduced into Oregon, and other parts of our country, by dispersal from fur ranches. They prefer the same fresh water systems as does our indigenous beaver, with which they compete for territory and food sources. They feed on vegetation near the water and can breed two

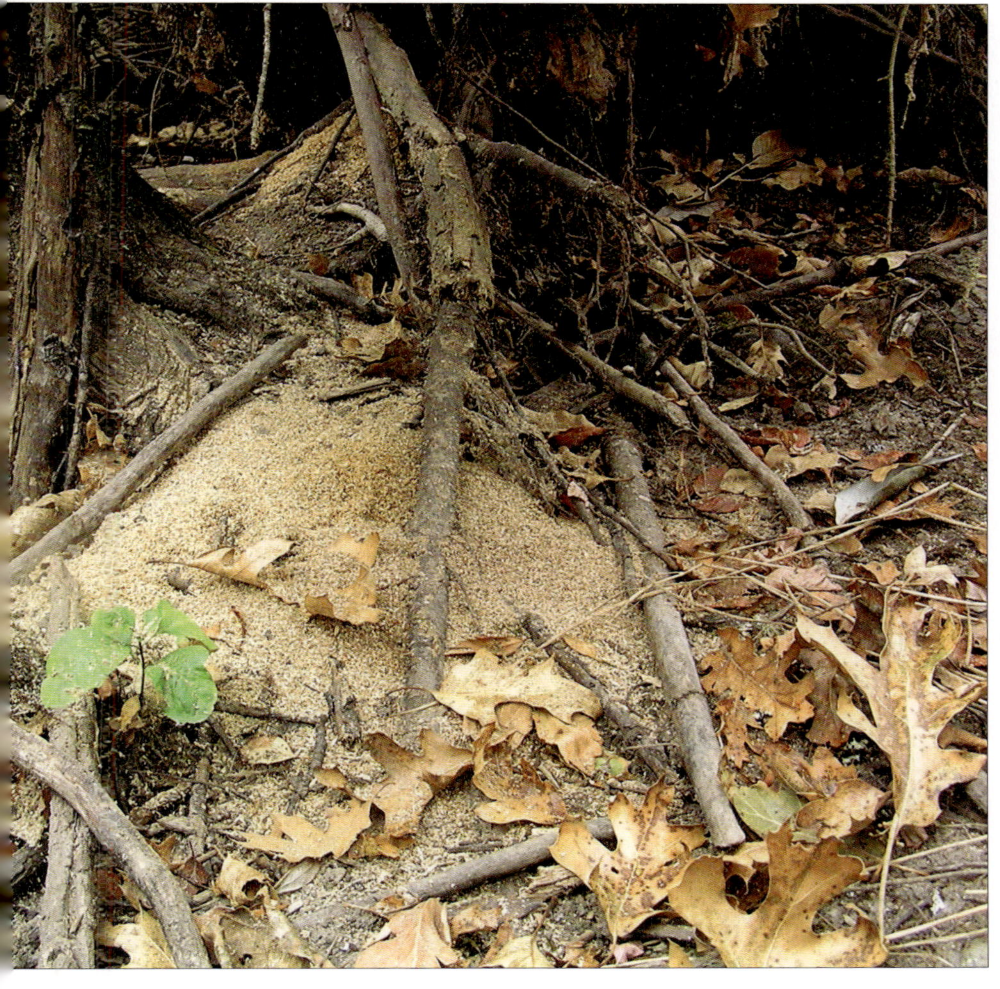

or three times a year with an average litter size of five. The beaver breeds once a year with two to four young born. Beavers are very beneficial and are ultimate wetland and riparian conservationists. What about nutria?

Continuing to look for mammal signs, a flash from a real animal runs across my visual field. I attempt to follow it, but to no avail. The mammal is a brush rabbit (*Sylvilagus bachmani*) that disappeared into its brushy home environment. I wait, just in case it may reappear, but no such luck.

This species can be active during the day, but most of their activities begin at dusk, extending throughout the night and into early morning. Although they have small home ranges of 70 to 300 feet in diameter (one source suggests they may be up to an acre for males, a half acre for females), their habitat lies in dense brush which provides cover for concealment from predators. I can attest to the difficulty of trying to walk and crawl around in this kind of habitat. Brush rabbits feed on various herbaceous plants throughout the year; but during winter, they add leaves and twigs of woody plants to their diet, and even needles and small twigs of Douglas fir that have blown down to their level by the winds.

Brush rabbits can breed up to five times a year with a litter of one to six young. Although this sounds like a lot of rabbits, they are prey for several different carnivores and some raptors; thus their prolific reproduction ensures survival of the species. They are known to be a favorite food item of the bobcat (*Lynx rufus*), which also inhabits the Preserve. And guess what. The bobcat's favorite habitat is brushy areas.

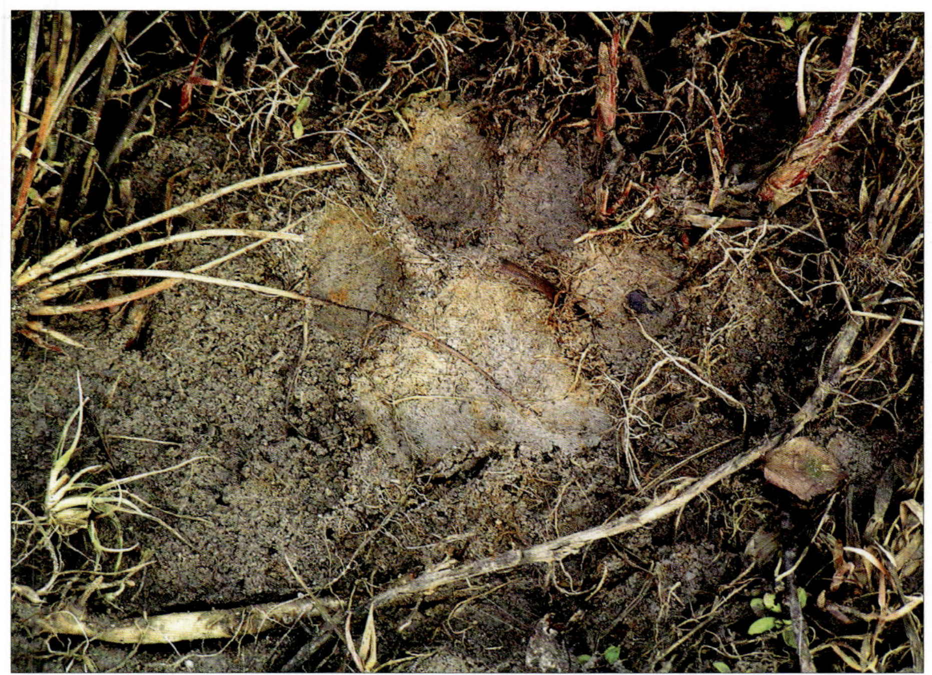

Occasional bobcat tracks and scat have been found during my investigations. They, along with other feline species, lead a highly predatory lifestyle, which is characterized by their dentition. Their teeth are designed for seizing and cutting their prey. In fact, many carnivores have a set of cheek teeth shaped like long shearing blades called carnassials. Carnivores' tight jaw attachments to the skull allow for up and down jaw movement, which accentuates the cutting function of these special teeth.

Bobcats, as do most predatory mammalian species, have forward-facing eyes, giving them binocular vision. This enables them to judge the distance between themselves and their prey before pouncing on it. Their highly specialized eyes assist them in hunting very effectively at night when most of their prey is active. Bobcats also have a good sense of hearing and possess sensitive whiskers, called vibrissae, that help them feel their surroundings. Whether this species is ambushing or stalking its prey, it will feed not only on brush rabbits but also on almost any other species that can be caught, including our problem voles.

The home range of this species is variable in our country, but where they occur in the west is about 7 square miles for females and 16 square miles for males. Although a male's home range may overlap other males, they definitely overlap those of females. As with all species whose male home ranges intersect those of multiple females, this overlap pattern importantly promotes genetic diversity within the species. For this solitary species, individuals avoid each other where there is home range overlap except during their annual breeding season.

As I leave the area where the fast-moving brush rabbit was last seen, I think about the balance of Nature and the biodiverse complexity of her faunal species, each of which evolved into an intricate design with mechanisms of adaptability to its lifestyle and to its environment.

* * *

The heat of summer begins to lessen as my favorite time of year, autumn, emerges in this cycle of life. I love watching Mother Nature engage in the creation of this special season with her pallet of brilliant colors. *"You have a big job in store for you,"* I speak to her in a giddy fashion. A light breeze kisses my cheek. *"I would love to help you if I could. I have some water colors at home."* I continue my side of the conversation, but with no response. So, instead, I sit within the roots of a tree and feel Nature's passion as

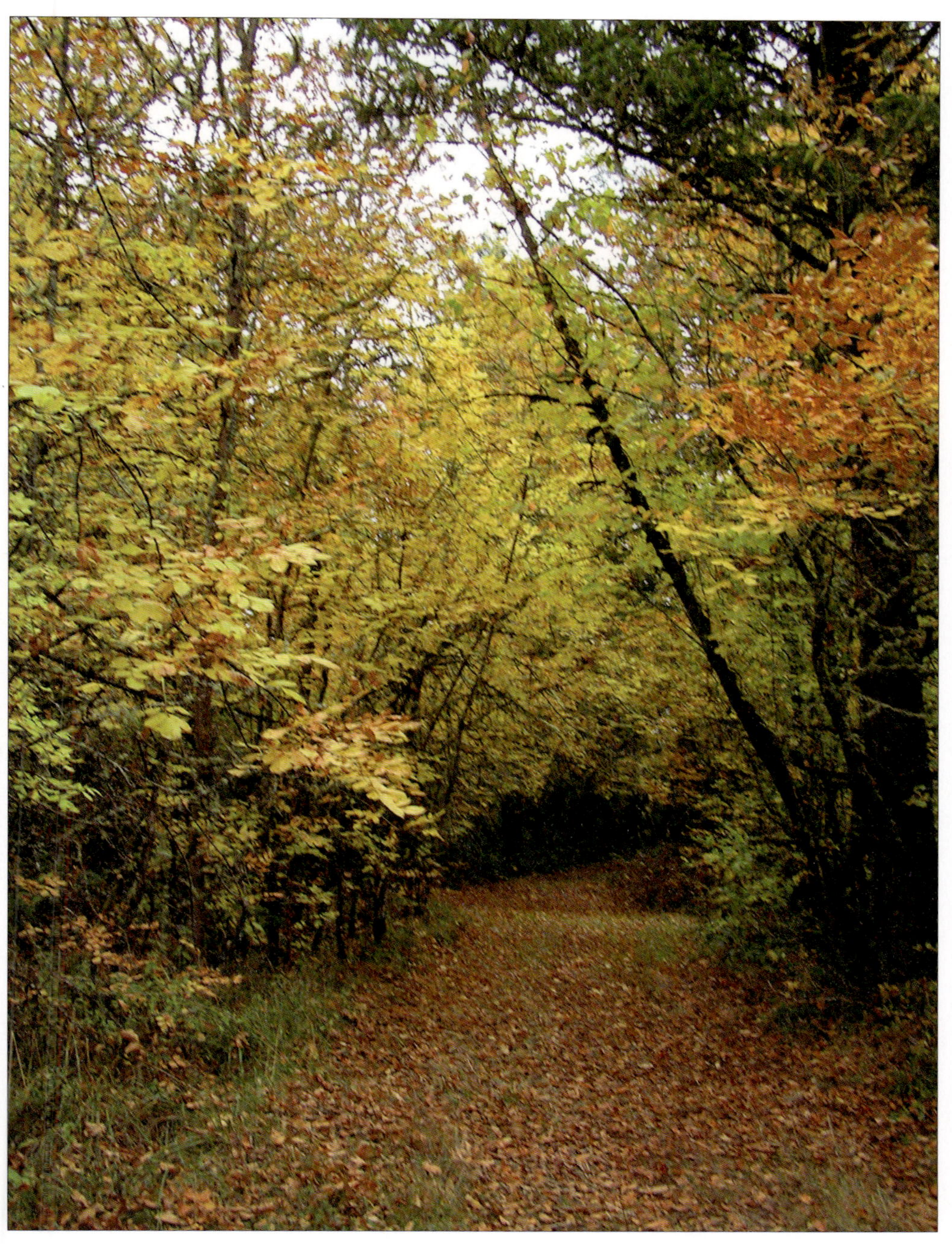

I watch the colors spread across her canvas. The entire setting becomes alive with tree leaves of color and golden grasses singing within her work of art. *"Thanks,"* I express to Mother Nature. While in an exhilarated state, I dance across this stage of artistic achievement.

It is time to return to my work, but in a high-spirited state. Ah, there goes a vole into a hole to possibly meet a mole underground with a gopher tunneling through. Let us enter their underground world and learn some of the fascinations of those seldom seen on the surface.

Pocket gophers are found throughout our country where the ground is suitable for digging. Different species show up in different locations, depending on their required habitat. They are fossorial, meaning they spend most of their life underground. Gophers are vegetarian, feeding on the roots of various plants, bulbs, and tubers. When they do appear above ground, they feed on nearby vegetation. About the only thing most people know about this group of animals is that they create mounds that "mess up" pristine lawns or ploughed fields, and eat some of the flora. What people don't realize is that gophers are very interesting and beneficial mammals, as we'll discover while moving up Mother Nature's ladder, beginning on the underground step.

Throughout my survey, as I encounter mole and gopher mounds, I become fascinated by the animals I will probably see only as specimens. I desire to learn more about those species that created these above-ground sculptures.

Let's begin with the gopher found in this Preserve, the camas pocket gopher (*Thomomys bubivorus*), an IUCN Red Listed Threatened Species. They only occur in the Willamette Valley of northwestern Oregon. This species, like other pocket gophers, is a solitary mammal; individuals come together only during their annual breeding season in the spring.

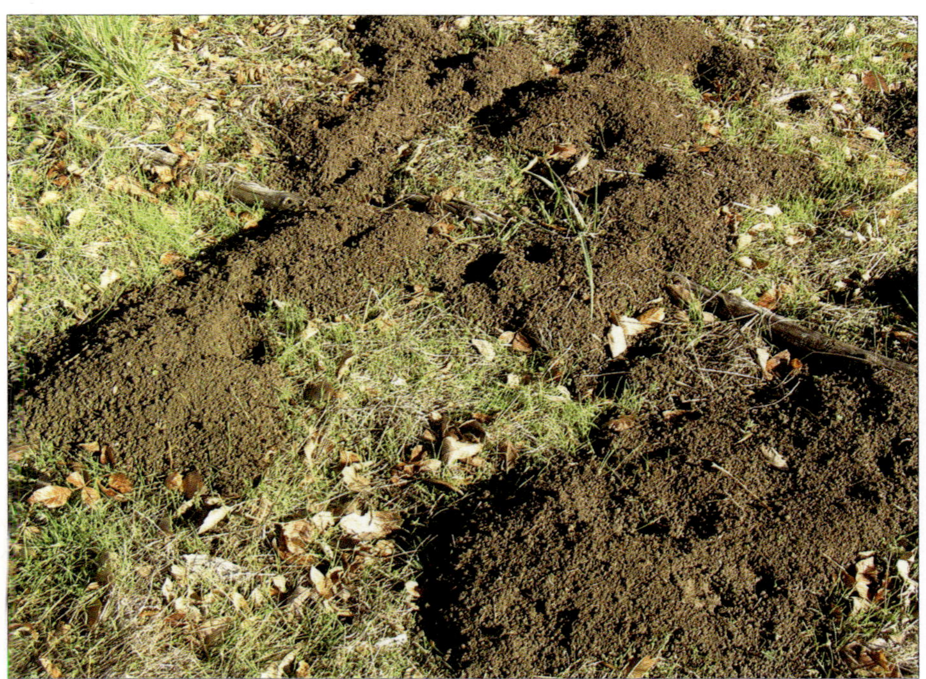

Gophers make fan-shaped mounds with an origin at their burrow's entrance from which the dirt radiates during their digging behavior. If they are not actively excavating the earth, they plug their entry with soil. These loose-fitting plugs provide a form of air-conditioning inside their systems and protection against unwanted visitors entering their homes "by closing the door."

Along with the colorful season of fall foliage, comes the beginning of the rainy season. As the moisture increases so does the number of surface mounds in the Preserve.

Research has shown that in the case of the camas gopher, when the soils are wet, this species increases its mound building. Interestingly, its larger mounds also have an opening on top — almost like an inverted chimney. This becomes increasingly apparent to me as I spend more time investigating their surface structures. It is hypothesized that the purpose for these openings is to increase ventilation and promote drying of their massive underground tunnels.

During the rainy season, I discover several smaller mounds alongside one very large mound that has a length of 66 inches and is 44 inches across at its widest point! Since not all camas gopher mounds are this large, this is a sight to behold. I stand in awe at the discovery.

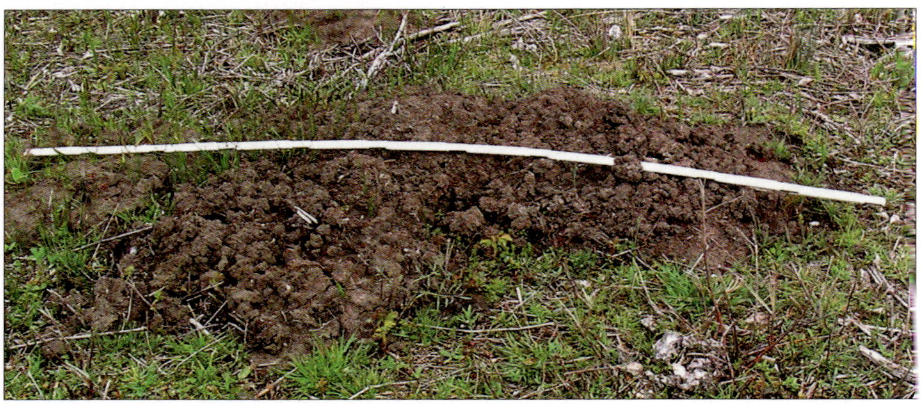

This gopher species is one of the largest — 12 inches in length and weighing in at about one pound. However, unlike most other gophers, they have weak claws for digging their underground tunnels, so instead they use their protruding incisors to loosen hard soil.

Gophers make two kinds of tunnels: shallow tunnels for gathering food; and deep tunnels for shelter, storing food, toilets, and nesting chambers. Interestingly, our grey-tailed voles are known to use the abandoned tunnels of camas gophers.

So . . . how do gophers' benefits mitigate their annoyances? Their tunneling behavior helps keep the earth porous. When the animals move away, these tunnels become a storage vessel for rainwater, thus precluding excessive surface runoff. They bury vegetation, which enriches the soil by forming humus. Also, seedlings from their food storage eventually work their way to the surface to become new vegetation. These micro-sites formulate the colonization of new plants from the species of flora upon which the gopher had previously fed. The cycle of life is never simple, but ever wondrous.

Continuing on the path of discovery, I encounter another solitary fossorial species that creates surface mounds — and at all times of the year. This animal is a mole which, unlike the gopher, is an insectivore. Instead of feeding on vegetation, their diet consists primarily of insects, insect larvae, earthworms, and other invertebrates.

Mounds of moles are often described as eruptions. They are referred to as "molehills" because they show no indication of an entrance into their surface structure. Moles work underground to create a deep tunnel; when pushing soil out through this tunnel, a surface structure is formed. They use their deeper tunnels most frequently, but also create surface tunnels for foraging or when trying to stay ahead of rising waters by moving in an upward contour of the land. This behavior has been shown to be preferred over exposing themselves on the surface of the ground. While using their front paws to excavate, they push the soil back with their rear feet.

As I discover some of the moles' more recently excavated tunnels, I observe the two-toned surface soil of

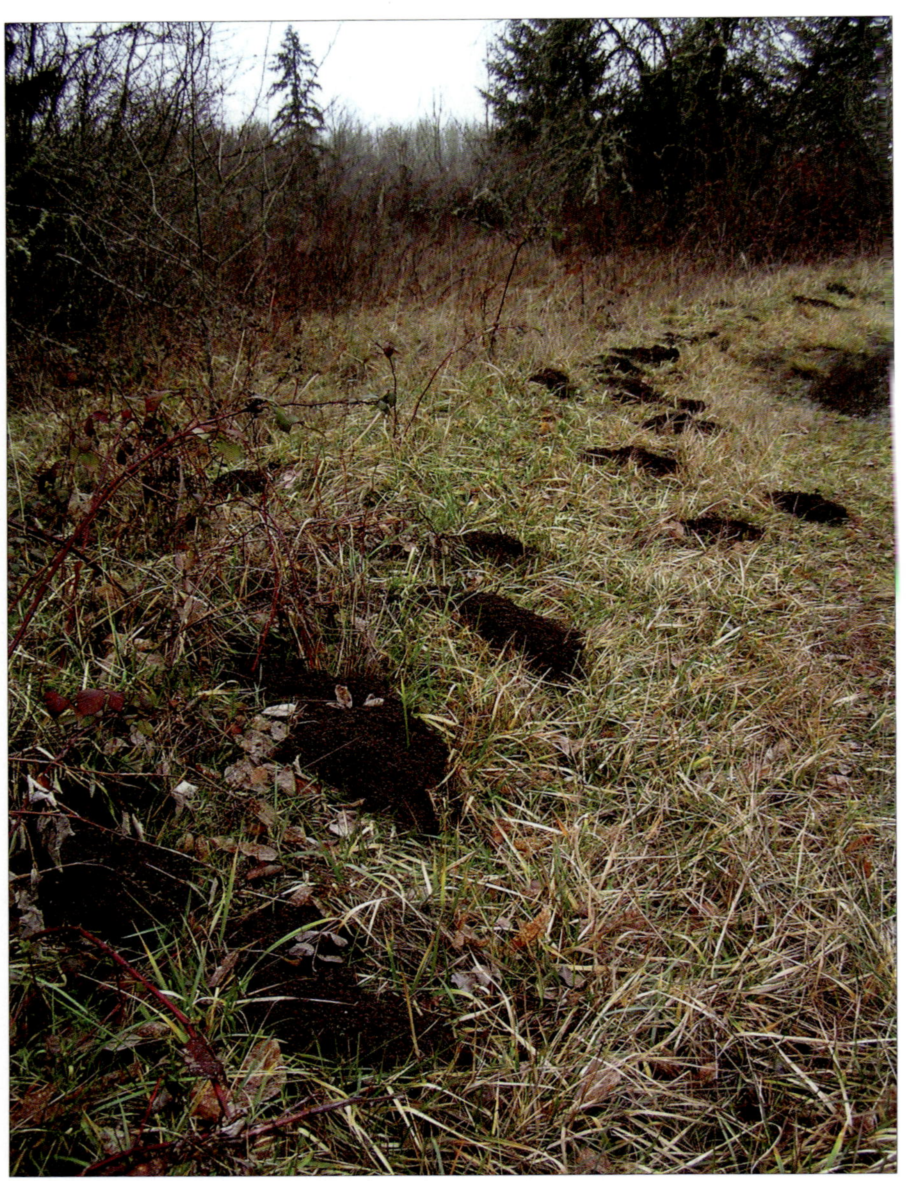

the exposed mounds. This is a clue to the depth at which they are digging. Their deep tunnels can extend downward to 10 feet, depending on the soil conditions. Since I am admittedly not a gardener, just trying to plant bulbs a few inches into the ground using a tool is hard work in our clay soil. So, sitting next to these mounds, I feel an appreciation for the work these animals must go through without human-made tools. Instead they use their large broad flat claws that fortunately evolved for heavy digging.

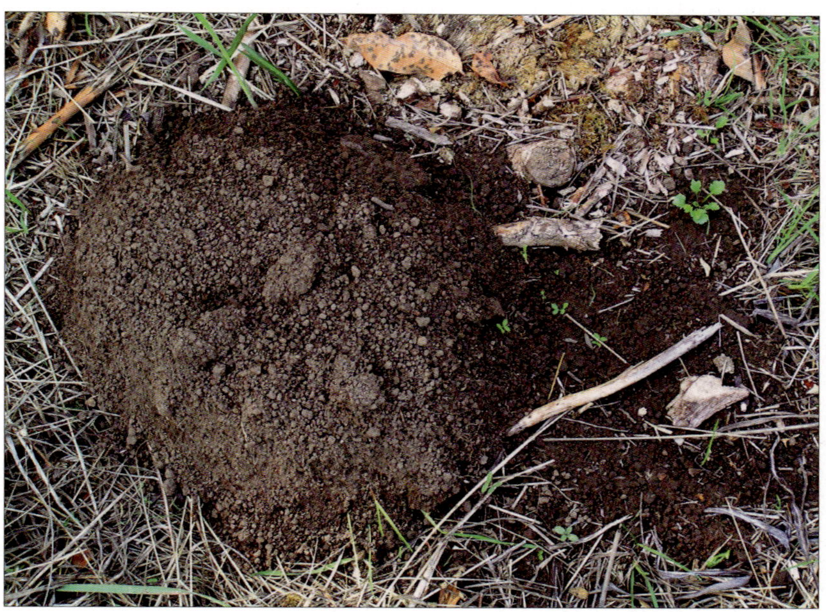

Our Preserve species is the larger Townsend mole (*Scapanus townsendii*) at about 9 inches in length and up to 6 ounces in weight. One account related to this species states that they are known to burrow into hard soil in seconds. Sitting in an area of mole eruptions, I try to dig a hole into the ground using my finger nails. Seconds turn into minutes until I give up the process.

Our mole species breeds once a year with two to four young born. Their nursery nests are often in slightly elevated land areas because they breed during the rainy season; these underground nests occur either near or under a larger mound of earth called a "fortress." Grasses may be pulled down through the soil to create cozy little nests.

This interesting species is also beneficial. Their tunneling behavior aerates and mixes soil layers. Additionally, they eat a significant number of insects and other invertebrate pests.

Abandoned mounds of both gophers and moles are used by other small mammals and are a critical retreat for certain reptiles and amphibians. Although humans may not have a high regard for some of these fossorial species, they do play an important role in the web of life.

* * *

The rains continue and so does the rise in water for our seasonal creeks. Enter one of my favorite mammals, the beaver. Because water does not occur in these streams during all months of the year, beavers that enter Willow Creek during the wet season are likely dispersed young searching for their own territories, which would make them about two years of age. This is the time when they reach sexual maturity. These beavers probably enter the Preserve from Amazon Creek, which flows throughout the year. Whatever the case, it is a time for jubilation.

From records known in Oregon, beavers were scarce in the Willamette River and its watershed as far back as 1824 due to over-trapping. During the early 1900s their numbers began to increase as the consequence of rigid legal protection. Now they are found throughout suitable waterways of Western Oregon and other parts of North America.

Beaver signs appear throughout my project. My tactile contact with the product of their diligent work bestowed upon me a special spiritual connection to the animal. I admit this unworldly feeling has occurred with other species of animals, during my many years of research, out of my love and respect for them and their role in the web of life.

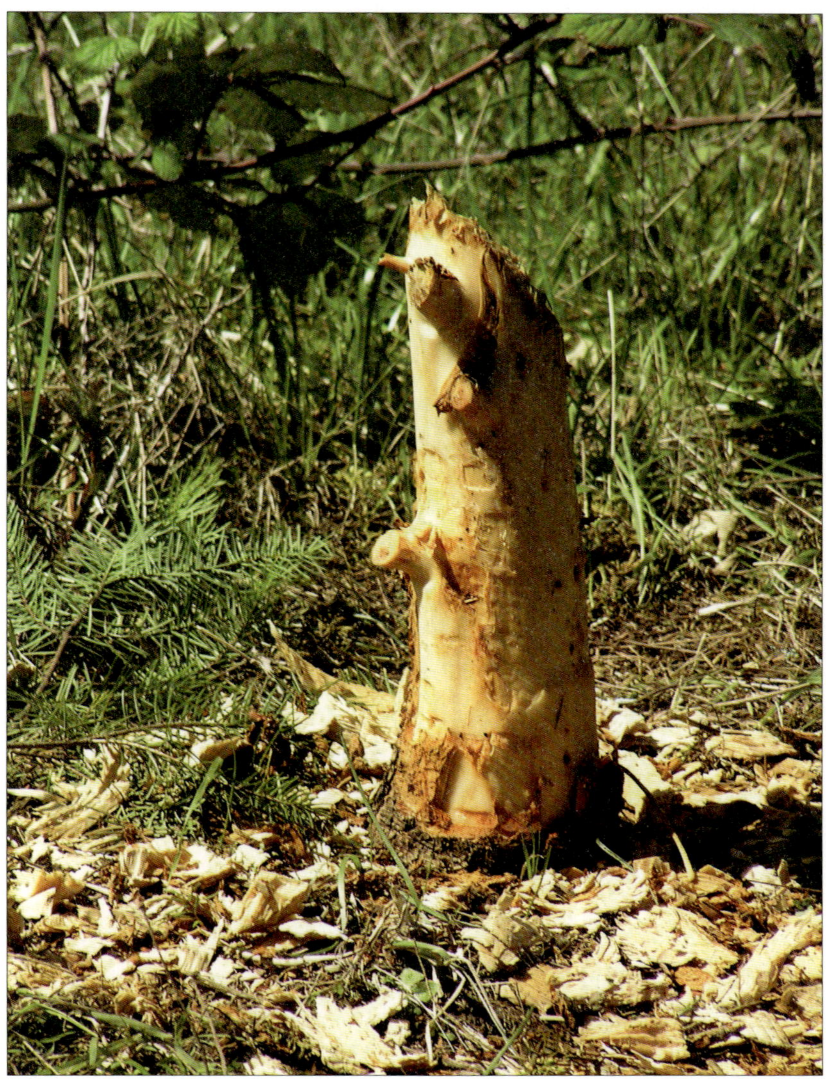

But there are a few species for which this depth of emotion is stronger; this is what I feel for the beaver. As someone once told me in relation to another spiritually connected animal, the river otter (as related in my earlier book *The Otter Spirit*), I brought, and continue to bring together, zoology and spirituality as one.

"Chip away, Mr. Beaver, at the chosen tree which provides your necessary nourishment and dam building materials."

When I started my project during the rainy season, there was one tree in particular that I watched disappear. It was across a waterway, so I could not reach it until later in the year, but I photographed it being recycled into a different layer of life. I did not see the full tree, but just its stump, so I imagine most of it became construction material for a nearby dam. Chips on the ground had distinct beaver tooth marks. This species will feed on the bark and cambium layer of deciduous trees, which is what must have occurred in this case. The leaves and twigs of this tree may have been foraged upon before I arrived on the scene.

One researcher refers to the beaver as a "choosy generalist herbivore." This species selects trees based on size, distance to travel, nutritional value, and palatability. They will also sample trees; odor and taste are found to be more sensitive senses than visual. Beavers also feed on leaves, buds, twigs, roots, and fruits of other deciduous woody plants, including acorns. They also feed on aquatic vegetation, riparian forbs and grasses. Beavers can rightfully be referred to as a choosy generalist.

I look around at all the trees and other vegetation on which the beaver could feed in this particular riparian

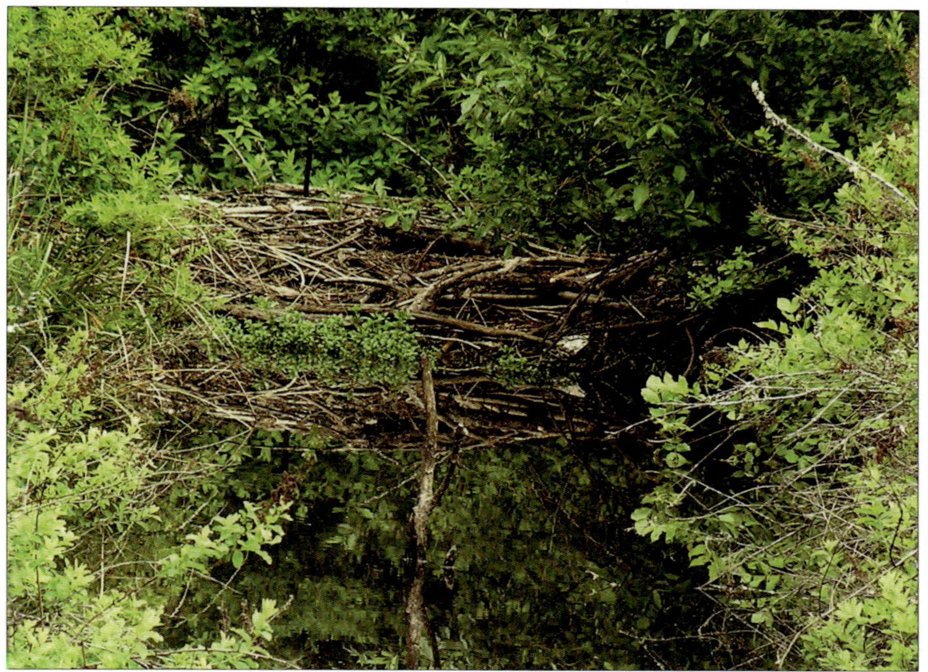

area. It's what I call a beaver's super market. In fact, felling some of these trees enables brushy vegetation to grow, thus providing habitat for brush rabbits and bobcats, not to say also a food source for our deer species. Additionally, through their dam building behavior, an area can have an increase in water level. If there are any non-felled trees in the area that can't withstand long periods of increased water, they may die. They then become snags which, in turn, provide habitat to various insects, birds, and mammals. I appreciate beavers' role in the web of life, including within this one small segment of Nature. Of course, there are many benefits of beaver behavior across the country, which were discussed in the previous story.

However, before we leave the beaver — which happens to be the Oregon state mammal — I share one other benefit of this species' behavior, and that is to Oregon's beloved salmon. The dams of beavers residing in coastal streams of our state can wash out during high water flows. This corresponds to the season when salmon are moving in from the ocean and swimming up towards a river's headwaters for spawning. In time, as the high flow into these coastal streams decreases, beavers respond by repairing their dams. These nutrient rich ponds and resultant cover creates a place for young salmon to thrive before they, in turn, set out to sea.

Continuing to check in this beaver-dam affected area, I discover some very familiar mammal signs from another semi-aquatic species. About two feet up from the water's edge, located on a raised mound of grasses, I find a "gift." From my prior research, the size, shape, and fishy odor of this scat lets me know it was left by a river otter (*Lontra canadensis*). Crayfish parts comprise a major portion of its composition, which corresponds to the discovery of this species residing in the creek. What amazes me is that the mobile otter also knew crayfish resided here. However it did, I was able to assess that behaviorally it took a roll in the grasses (determined from signs) to dry its thick fur before moving onward to a place only it knows.

The river otter is often referred to as the joyful, playful clown of the animal world. It can slide in the snow, or, like it does here, on slick mud banks. But not only are otters a charismatic species, they are also a "flagship species" to water quality. Because otters' health is susceptible to the effects of pollution, they are considered a bio-indicator for the environmental health of fresh-water habitats. Another benefit of this species is that their aquatically derived nutrients (primarily from fish) are deposited (through scat) onto land. Research has shown that their deposits fertilize terrestrial vegetation, which, in turn, influences the growth and prevalence of particular plant species important to riparian and wetland habitats. Then this nutrient transference onto land eventually works its way back into fresh waters through reverse flow patterns. So not only do otters bring us joy, but they also play an important role in maintaining healthy fresh water systems.

Two other species in the same taxonomic family as otters are also found in the Preserve, mink (*Mustela vison*) and long-tailed weasels (*Mustela frenata*).

Mink, like river otters, are semi-aquatic. They are an opportunistic carnivore that feeds on a variety of prey, especially aquatic and semi-aquatic species. However, research has shown that they also feed on voles from grassy areas surrounding locations of fresh water systems, like we have in the Preserve. I discovered some dens embedded within banks of one of the creeks which would be the correct size for mink. Since they are known to occur in nearby Amazon Creek, those who were in the Preserve would have moved on when Willow Creek dried up. Young mink disperse during autumn, so if they were here, they, like the beavers, were probably dispersing young. I also discovered some scat during the wet season in a location near the bank dens. This scat was comparable to mink droppings found during my prior research in the Rockies.

Long-tailed weasels prefer grassy areas near water, where they have been seen in the Preserve. Their prey is primarily rodents and other small mammals. Because of their flexible, long slender body, they can follow a mouse into its burrow, and then take over the burrow. This species can be as long as 22 inches and weigh up to 11 ounces. In snow country, their pelage molts from shades of brown to camouflage white in winter.

At least one manner that a weasel captures a rodent is to throw its body in a loose, snake-like coil over the prey's body, subduing it before gripping the back of the head for the killing bite. Their predatory behavior of feeding on voles and other small rodents, such as our deer mice (*Peromyscus maniculatus*), is also beneficial to keeping life in balance.

* * *

The wet season rains are beginning to subside as our senses are treated to the onset of spring. One of the harbingers of this special season is the flowers of "pussy" willows. Now the end of March, I sit near a beaver dam where a breeze causes the small flowers to dance like fairies in the glistening sun. Those blown into the waters look like small aliens from an unknown world. Some of the remaining attached flowers will soon turn into fruits whose tiny cottony seeds will then be distributed by wind and water. But these seeds must eventually land on moist soil in order to begin their life-giving process.

A small sparrow enters this scene flittering along the bank before landing to scratch with its feet within vegetation in search of a meal. This small bird and these tiny seeds take up so little space

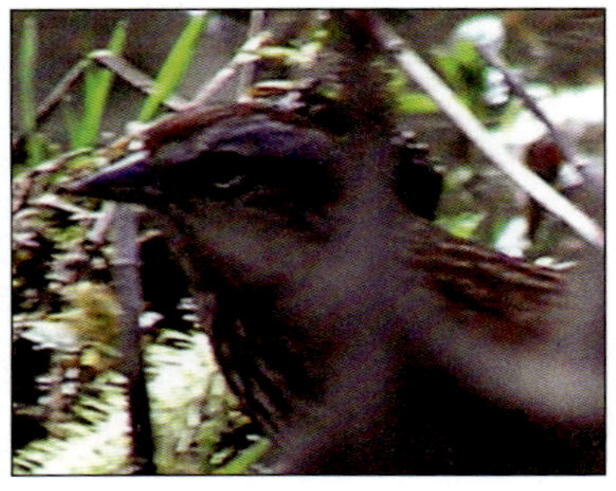

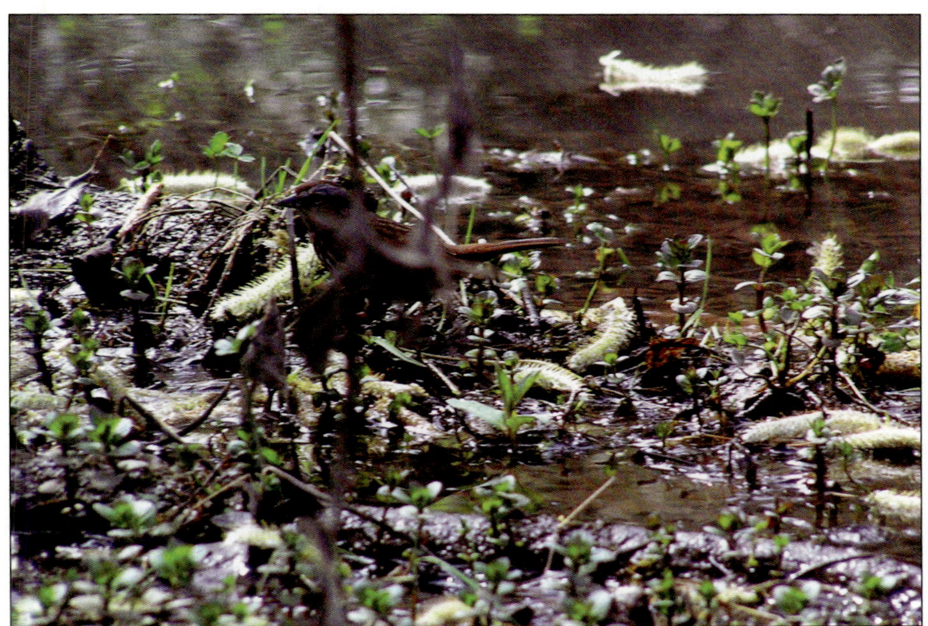

and weigh almost nothing, yet each species plays its role in the complex world of Nature.

I pick up a branch whose outer covering has been removed, and, at its end, are the teeth marks of my friend, the beaver. Beavers are a beneficiary of willows but also a contributor to their continued growth, as told in the previous story.

I must try to climb back up the muddy slope down which I slid, like an otter, to the water's edge. Although my slide was accidental and not in play, I'm glad it occurred so I could enjoy this Nature moment.

* * *

Did I or didn't I just see a small pointed-nosed animal scurry across my path? Not to make fun of it, but this animal's long narrow snout makes it look like a little witch.

Hoping to get a better look, I try to follow its path of travel. Although hopeless, I do know there are at least two species of this seldom seen group known as shrews.

Shrews are in the same taxonomic order, *Insectivora*, as our previously discussed moles, and also have a similarly shaped snout. Shrews are not a fossorial species like our moles. They feed primarily on insect larvae, compared to the more varied diet of moles. Since we have two different kinds of shrew in the Preserve (*Sorex vagrans*) and (*Sorex pacificus*), partitioning is at work. They not only prefer different habitats but also feed on different larvae; one species feeds on those of terrestrial origin and the other feeds on those of aquatic origin.

Progressing along my path of travel, I see some tracks ahead which are like those I have discovered throughout

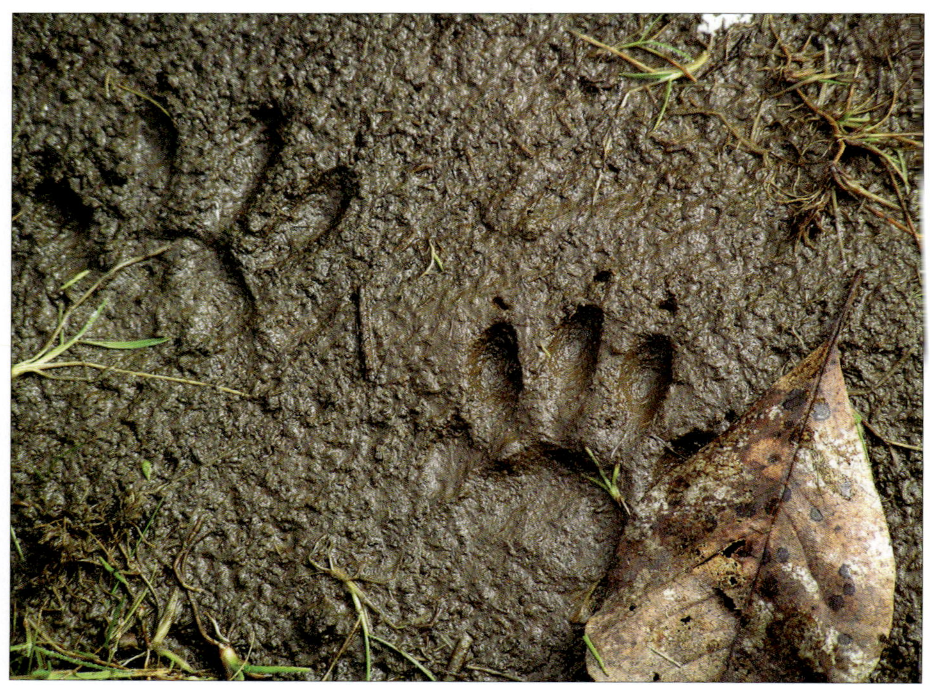

all months of the year in this Preserve. Because the animal that produces these familiar tracks has a black mask across its eyes, it is often referred to as a "bandit." I refer to it as the familiar raccoon (*Procyon lotor*). The tracks I discovered over the months look like a small human hand. This corresponds to the shape of their paws and their mode of terrestrial movement — flat-footed on naked soles.

Their hands and feet offer them much dexterity; they can climb trees, and even open latches or garbage cans like a bandit. Raccoons are omnivores; they feed on both plants and animals. Their tooth structure consists of sharp slicing canines for eating meat and broad flat molars for eating vegetation. They are opportunistic and will feed on anything they can find or catch, whether on the ground or in trees. In addition to preying on various animal species, in our area they feed on seasonal nuts, acorns, berries, and apples.

It is often thought that raccoons wash their food prior to eating. Actually, they have a well-developed sense of touch in their forepaws. Research has shown that this species has a large number of nerves associated with touch in their forepaws and a high degree of developed areas in their brain associated with touch. Since they pick up food with their hands, moving it between their paws may actually give them tactile pleasure. Even without access to water, they are still observed "washing" their food between their paws.

What to my wandering eyes does appear? Cautiously, I approach the animal but it doesn't move. So is it dead or "playing possum"?

A closer look tells me it has become a specimen. The Virginia opossum (*Didelphis virginiana*) is our country's only native pouched mammal, or marsupial. Actually, the death-feigning behavior of this species is a passive-defensive tactic. The idea is that this behavior may cause a pursuing predator to lose its visual cue of motion and walk away. Another defense used by opossums during this behavior is emitting odorous anal secretions, which may also make them appear distasteful to potential predators.

Opossums moved into North America from South America 3.5 million years ago via the Pliocene land bridge. Between natural dispersal and introductions over the years, it has become one of the most adaptable and prolific species of our country. Interestingly, their ancestors are a member of the oldest known taxonomic family from the Cretaceous Geologic Period, about 100 to 70 million years ago. Although our species is from a more recently evolved genus, it is still a structurally primitive life form which has changed very little from its early ancestors.

Slow-moving opossums are both at home in trees and on the ground. Like our raccoon, they are omnivores; thus, they have a varied plant and animal diet. Cheek teeth (molars and premolars) in mammals are used for crushing and grinding food prior to digestion. In the opossum, these teeth are more pointed and sharp compared to those of raccoons, but each species has nicely adapted to its particular diet. Looking at the long, sharp canines of each in my home laboratory tells me neither would have a problem capturing its animal prey.

Interestingly, raccoons and Virginia opossums co-exist throughout much of their range. Their foraging patterns appear to be independent of the other, even though their periods of major activity lie between sunset and sunrise.

Onward we travel to see what mysteries Mother Nature will present on her stage of life as winter gives way to spring. The work of a Nature detective never ceases.

* * *

Spring is the blossoming of the bud into the blooms of life. Color spreads across the landscape from a variety of floral species. However, there is one in particular that stands out above the rest: the beautiful blue-to-purple flowers of the camas. On occasion, I have even found a white-flowered camas amongst the sea of blue. These species particularly

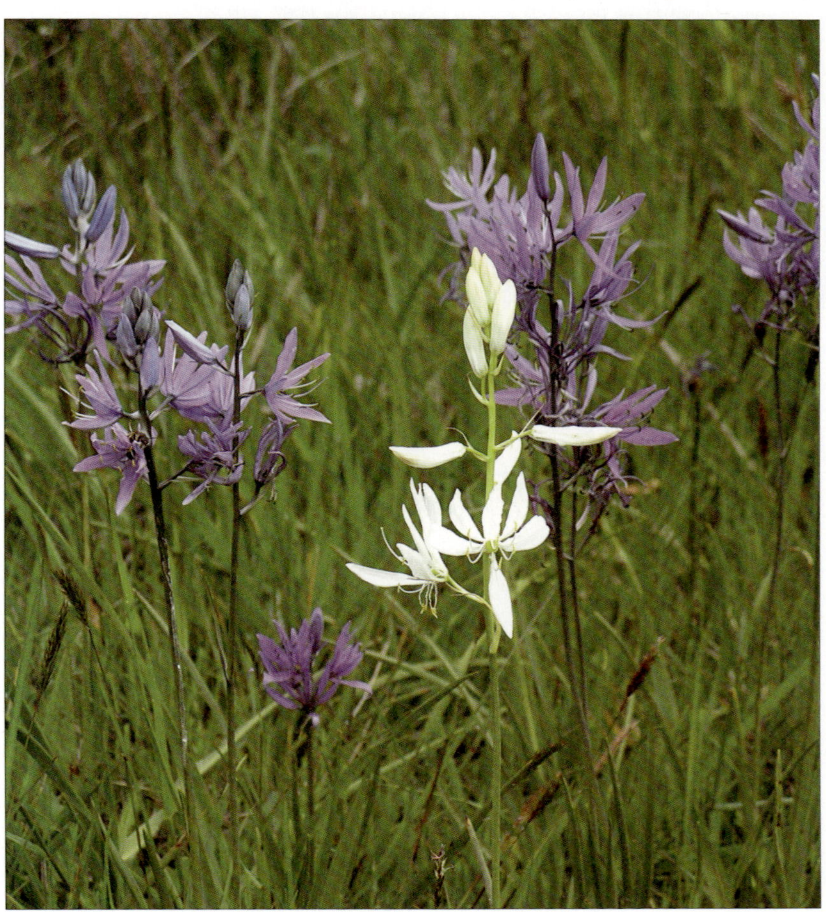

love the moist meadows of the Preserve, which is their habitat wherever they occur.

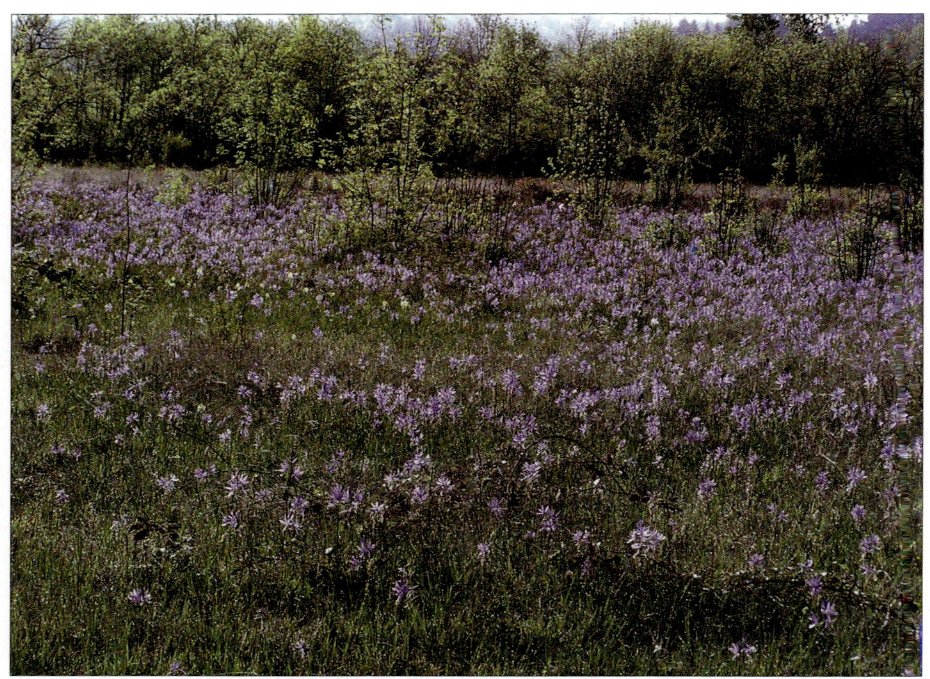

Before the European invasion into the Pacific Northwest, people from indigenous cultures, particularly those of the Columbia Plateau, used camas bulbs as a vegetable staple. They dug up and ate limited quantities in early spring, possibly to help with their nourishment after a lean winter. Then, when the plants had nearly finished blooming (June and July), the heaviest harvest occurred. Camas bulbs were eaten either raw or cooked. If they weren't eaten immediately, the people processed them to ensure they would remain edible for a long period of time, particularly storable as a winter food source.

Remember the camas gopher? They may have been given their common name because of a fondness for camas bulbs, but apparently that isn't known for sure. Their species' Latin name (*bulbivorus*) means "bulb" and "to devour." Because they feed on a variety of bulbs, tubers, and other vegetation (not just camas bulbs), their impact on this floral species in the Preserve is either slight or nonexistent, as I look out onto a beautiful carpet of undisturbed camas.

Amidst the floral splendor which surrounds us emerges the melodic song from the bird with a golden voice, the lovely meadowlark, Oregon's state bird. Its song springs forth with such clarity and beauty that even the sun emits a warm smile.

Many years ago, I conducted a project on meadowlarks roosting in a salt water marsh of southern California. Because of the Pacific Ocean's influence, I very often arrived at my study site in a state of fog, literally. However, when the weather was right and the birds cooperated, it was a nice environment to spend time.

Meadowlarks are usually found in drier habitats such as open fields, meadows, and grasslands. When they feed, they walk on the ground rather than hop like some passerine species. Their diet consists primarily of insects, with a smaller percentage being grains and seeds. Since we are now into spring, males are setting up their territories prior to the arrival of females. When the females arrive and the two sexes come together, courtship begins. Their nests are constructed on the ground, composed primarily of grasses, often partially roofed and with runways.

After their spring-summer breeding season and late summer molt has occurred, meadowlarks form flocks that last until the next breeding season. I discovered during my project that the flocks form in late September. Behaviorally, the flocks will come into their roost 15 to 30 minutes before sunset. Some singing and movement occurs until about 20 minutes after sunset when they settle on the ground in amongst the vegetation. Then, approximately 20 minutes prior to sunrise, the flocks leave the roost to resume diurnal behaviors. This gregarious behavior continues until April after which the birds separate to resume their breeding cycle. I can't say for sure that these behaviors occur throughout the Pacific Northwest, but it makes sense that would be the case.

Even though meadowlarks are not mammals, there is an interest in this beautiful state bird. Therefore, I decided a diversion was appropriate. From personal experience, when my late spring/summer early morning arrivals at the Preserve are greeted by this bird's flute-like song, I know a cheerful journey lies ahead.

Raccoons are a concern as a potential predator of ground-nesting birds such as meadowlarks. An interesting study conducted on a grassland reserve shows that raccoons avoid nest-searching behavior in this type of habitat during bird-nesting season. Instead, they move across grasslands to areas of richer food sources found in forested riparian and wetland habitats. Then during leaner times of fall and winter (after ground-nesting season ends), there is more searching for food in all habitats.

* * *

Today I walk off trail, moving into a forested location to sit on a stump and contemplate my project. After a few quiet moments, I return to the nearby trail to continue my journey. There, on my trail, almost on an even line with where I was sitting, is some nice fresh red fox (*Vulpes vulpes*) scat. Actually, I did check this area before moving into the forest and it was not here. Well, I guess the animal saw me, although I was not fortunate to see it. One researcher, who observed that they may see you but you don't see them, calls the red fox a "ghost of the forest." That is certainly the case with my experience.

Although this species occupies about every terrestrial habitat except deep forest, they prefer meadows interspersed with patches of brush and timber. Their home range can extend from ½ to 6 square miles under natural conditions, or 24 acres in suburban areas. Many factors are involved in the home ranges of our predators. Their size is dependent on type of habitat, prey availability, as well as natural and unnatural barriers. One long-term researcher found, across various studies, that coyotes' home ranges average about 8 square miles. However, they can be larger or smaller depending on ambient conditions.

The scats of red foxes and coyotes are often found where two trails cross. I discovered, in the case of both species, that their scats were found where a deer trail crossed the main trail. I sometimes found it in the middle of a trail.

Like the coyote, red fox are opportunistic feeders. Their main diet consists of rodents, rabbits, insects, and fruits. When hunting small rodents, such as voles and mice, they stand motionless, while using their visual and auditory senses. Once prey is detected, they leap, bringing their forelegs down to pin the animal. This is the same behavior I observed during winter at a snow bank in the Rockies.

Coyotes and red foxes co-exist over much of their range in North America, even though they feed on many of the same small prey. Interestingly, during summer, both species also become "fruitarians." I discovered seeds from fruits in the summer scat of both species. Some studies have shown there is a high degree of inter-specific tolerance between these two species; however, coyotes can be antagonistic towards red foxes when they are hunting in pairs.

Continuing on our journey, a mammal whose tracks I often see, and who physically presents itself to me on occasion, is the black-tailed deer (*Odocoileus hemionus columbianus*). I have been privileged to see their beautiful fawns appearing on grassy meadows, before running with mom into treed locations once I am detected. Their breeding season (rut) occurs in our area during November and December. Fawns are born about seven months later. Interestingly, members of their genus (*Odocoileus*) have metatarsal glands, which occur on this deer's hind leg or heel, visible as tuffs of long stiff hairs. Females and their young sniff these glands for recognition purposes.

Black-tailed deer are herbivores, which predominantly graze in summer, but usually browse during other times of the year. Their genus does not possess upper canine teeth so, instead of the upper and lower canines working together, they use their highly mobile lips and prehensile tongue to draw vegetation across their lower teeth. Then, they cut off the vegetation much like a tape dispenser cuts off tape. Since they are prevalent in our area, quietly watch them feed sometime.

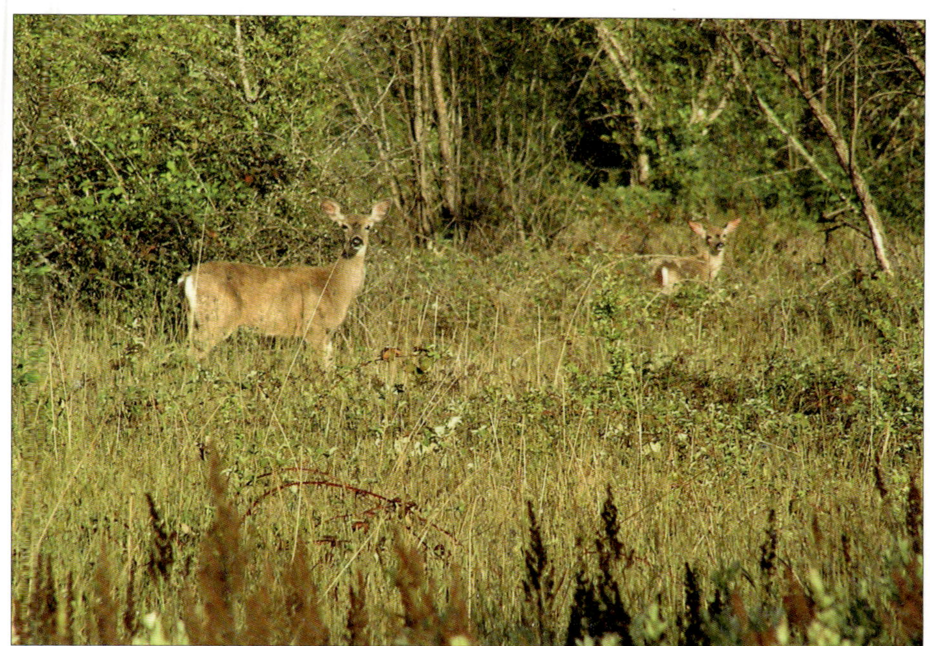

Deer are members of the suborder *Ruminantia*, which are even-toed, hoofed mammals with three or four-chambered stomachs that "chew their cud." How does their feeding process work? Basically, they swallow their food un-chewed. Then they regurgitate it and chew it thoroughly before re-swallowing it. This process allows them to quickly fill their larger stomach chamber while feeding in the open. When the animal retires to lie down under cover of grasses or forest, it continues the more lengthy process of chewing, thus reducing time exposed to predators.

Evolution and adaptation continue to amaze us.

* * *

Other mammal species reported by qualified people near the Preserve's office include western gray squirrel (*Sciurus griseus*), California ground squirrel (*Spermophilus beecheyi*), and a chipmunk (*Tamias sp.*). Even though I did not observe these animals or their signs, they are still an important part of my survey. Sometimes I found potential den sites in my major areas of concentration; but without seeing the animal or finding their signs near by, I could not determine to whom these homes might belong.

Even a black bear (*Ursus americanus*), observed in the fall near upper portions of the Preserve, probably entered the Preserve at some point in time. In fact, before my project began, one of the Preserve staff observed a nice clear set of bear tracks.

Another mammal whose tracks were observed before my project began is that of the mountain lion (*Felis concolor*). I also found a potential mountain lion track during my project on the side of a stream bed. It was definitely feline and too large to be from a bobcat.

I had previously observed black bear and mountain lion tracks during my projects in the Colorado Rockies. One time I discovered bear tracks along with a small bouquet of flowers. What was that message all about? Although I occasionally saw a black bear in the distance, I more often found their signs.

Unfortunately, I never saw mountain lions in the Rockies, even though one particular area was perfect habitat for them. Of course, this solitary secretive species has a large home range. The home range of a female may cover about 70 square miles; that of a male may extend up to a few hundred square miles, overlapping those of several females — again, to promote genetic diversity during breeding. Thinking back when I first started my mammal research, a colleague and I thought we could study mountain lions. We had a preserve chosen where a mountain lion had occasionally been observed. Although I have patience, knowing what I now know, I think that would have been a mistake. However, some day I hope to see this beautiful mammal moving silently through my own life cycle.

* * *

This mammal survey project in Willow Creek Preserve offered me an opportunity to become an intimate part of a bio-diverse array of fauna residing within both varied and specialized plant communities. Throughout my prior thirty-three years of mammal research, I primarily concentrated on one species at a time. This project was a first for me and I am appreciative of ending my career on such a positive note.

I learned a lot about an amazing number of species and feel fortunate to share this information with you, the reader. Now, you should journey into Nature and become a detective yourself to see what you can discover. More mysteries abound in the Natural World than the ones I have experienced.

* * *

My final message to you, the reader, is to never take Nature for granted. Each of her species of fauna and flora has evolved and adapted to become an important part of the layered web of life.

For the human species to continue to exist in this world, respecting, protecting, and preserving other species of life is the only answer. We already see what is happening on

Earth when that is not the case. Disrupting Nature's flow of life is also leading to the destruction of human kind.

Maybe, you are the one to turn around the tide of destruction. Maybe, you are the one to make the difference. PLEASE TRY! Mother Nature is trying to guide you . . . .

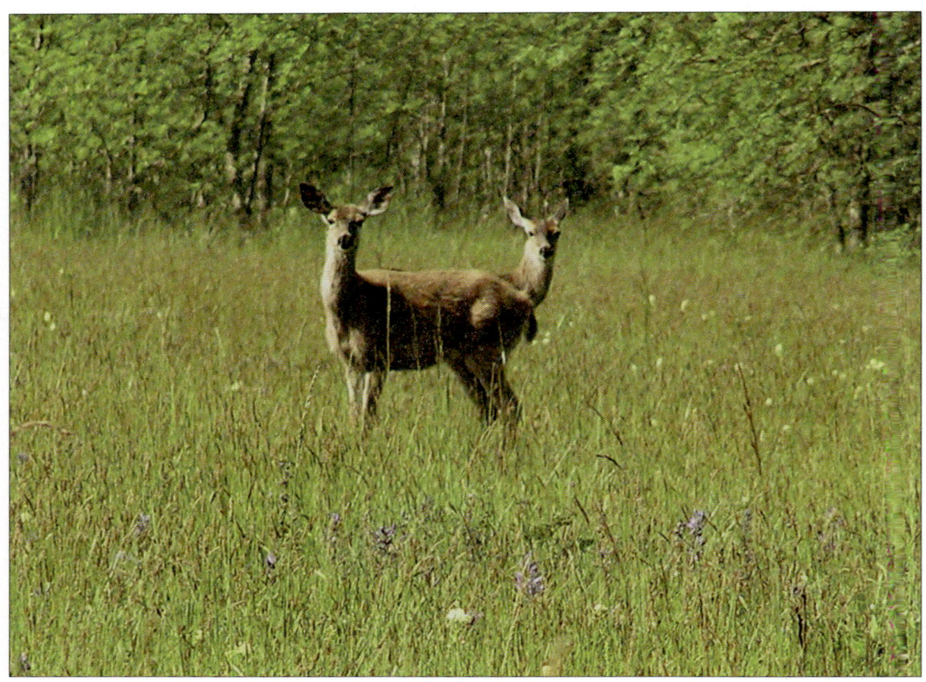

References

Bailey, V. 1936. *The Mammals and Life Zones of Oregon*. U.S. Department of Agriculture - Bureau of Biological Survey. Washington, D.C. North American Fauna 55: 1-416.

Bekoff, M. 1977, *Canis latrans. Mammalian Species* No. 79: 1-9.

Bekoff, M. 1982. Behavioral ecology of coyotes: social organization, rearing patterns, space use, and resource defense. *Z. Tierpsychol.* 60:281-305.

Ben-David, M., R.T. Bowyer, L.K. Duffy, D.D. Toby, and D.M. Schell. 1998. Social behavior and ecosystem processes: river otter latrines and nutrient dynamics of terrestrial vegetation. *Ecology* 79:2567-2571.

Berg, J.K. 1999. *Final report of the river otter research project on the upper Colorado River basin in and adjacent to Rocky Mountain National Park*, Co. National Park Service.

Berg, J.K. 2000. North American river otter diet. *River Otter Journal*. Vol.IX No. 2:4-5.

Berg, J.K. 2005. *The Otter Spirit: A natural history story*. Ulyssian Publications, WA. 1-l83.

Berg, J.K. 2010. *Final report of the mammal survey on Willow Creek Preserve in Eugene, OR*. The Nature Conservancy: Eugene, OR.

Blair, W. F., A.P. Blair, P. Brodkorb, F.R. Cagle, and G.A. Moore. 1968. *Vertebrates of the United States: 2nd Edition*. Section on Mammals: 454-562. McGraw-Hill Book Co.

Bontrager, D. 2008. Examination of skeletal material from carnivore scat breakdown in Willow Creek Preserve. Personal communication.

Borgnia, M., M.L. Galante, M.H. Cassini. 2000. Diet of the Coypu (Nutria, *Myocastor coypus*) in agro-systems of Argentinean Pampas. *J. of Wildlife Mgmt*. 64(2):354-361.

Bowen, W.D. 1982. Home range and spatial organization of coyotes in Jasper National Park, Alberta. *J. Wildlife Mgmt*. 46:201-216.

Boyd, S.K. and A.R. Blaustein. 1985. Familiarity and inbreeding avoidance in the gray-tailed vole (*Microtus canicaudus*). *J. of Mamm*. 66:348-352.

Brown, L.N. 1975. Ecological relationships and breeding on the nutria (*Myocastor coypus*) in Tampa, FL area. *J. of Mamm*. 56:928-930.

Carraway, L.N., L.R. Alexander, and B.J. Verts. 1993. *Scapanus townsendii*. *Mammalian Species* No. 434:1-7.

Carver, B.D., M.L. Kennedy, A.E. Houston, and S. B. Franklin. 20ll. Assessment of temporal partitioning in foraging patterns of syntopic Virginia opossums and raccoons. *J. of Mamm*. 92:124-139.

Chapman, J.A. 1971. Orientation and homing of the brush rabbit (*Sylvilagus bachmani*). *J. of Mamm*. 52:686-699.

Cornely, J.E. and B.J. Verts. 1988. *Microtus townsendii*. *Mammalian Species* No.325:1-9.

Crowley, S., J. Johnson, and D. Hodder. 2012. Spatial and behavioral scales of habitat selection and activity by river otters at latrine sites. *J. of Mamm*. 93: 170-182.

Cressman, L.S. 1981. *The Sandal and the Cave: the Indians of Oregon*. (Originally pub. in 1962 by Beaver Books). Oregon State U. Press. Corvallis.

Duffy, L.K., R.T. Bowyer, J.W. Testa and J.B. Faro. 1996. Acute phase proteins and cytokines in Alaskan mammals as markers of chronic exposure to environmental pollutants. *American Fisheries Society Symposia* 18:809-813.

Eisenberg, J.F. 1983. *The Mammalian Radiations*. U. of Chicago Press, IL.

Elbroch, M. 2003. *Mammal Tracks & Sign: A guide to North American species*. Stackpole Books: 1-780.

Giger, R.D. 1973. Movements and homing in Townsend's mole near Tillamook, Oregon. *J. of Mamm*. 54:648-659.

Goertz, J.W. 1964. Habitats of three Oregon voles. *Ecology* 45:846-848.

Gorman, J.L. and R.D. Stone 1990. *The Natural History of Moles*. Cornell U. Press, N.Y.

Grinder, M.I. and P.R. Krausman. 2001. Home range, habitat use, and nocturnal activity of coyotes in an urban environment. *J. of Wildlife Mgmt*. 65(4):887-898.

Grant, P.R. 1971. Experimental studies of comparative interactions in a two species system. *Microtus and Peromyscus* species in enclosures. *J. Anim. Eco*l. 40:323-350.

Hopkins, S.S.B. and E.B. Davis. 2009. Quantitative morphological proxies for fossoriality in small mammals. *J. of Mamm*. 90(6):1449-1460.

Kays, R.W. and D.E. Wilson. 2002. *Mammals of North America*. Princeton U. Press, Princeton: 1-140.

Koehler, G.M. and M.G. Hornocker. 1991. Seasonal resource use among mountain lions, bobcats, and coyotes. *J. of Mamm*. 72(2):391-396.

Krebs, C.J. 1996. Population cycles revisited. *J. of Mamm*. 77(1):8-24.

Kuhn, L.W., W.Q. Wick, and R.J. Pedersen. 1966. Breeding nests of Townsend's mole in Oregon. *J. of Mamm*. 47:239-249.

Lariviere, S. and L.R. Walton. 1998. *Lontra canadensis. Mammalian Species* 587:1-8.

Linzey, A.V. and G. Hammerson. 2008. *Thomomys bulbivorus*. In IUCN 2009 Red List of Threatened Species. Version 2009.2.

Loukmas, J.J. and R.S. Halbrook. 2001. A test of the mink habitat suitability index model for riverine systems. *Wildlife Soc. Bulletin*. 29(3):821-826.

Maser, C. 1998. *Mammals of the Pacific Northwest: from the coast to the high Cascades*. Oregon State University Press, Corvallis, OR.

Melquist, W.E. and A.E. Dronkert. 1987. River Otter. Pp:627-641, in *Wild furbearer management and conservation in North America* (M. Novak, J.A. Baker, M.E. Obbard, and B. Malloch eds.) Ontario Ministry of Natural Resources, Toronto.

Moore, A.W. 1939. Notes on the Townsend mole. *J. of Mamm*. 20:499-501.

Miller, F.W. 1931. A feeding habit of the long-tailed weasel. *J. of Mamm*. 12:164.

Murie, O.J. 1974. *Animal Tracks: Peterson Field Guides*. Second Edition. Houghton Mifflin Company, Boston: 1-375.

Newbury, R.K. and T.A. Nelson. 2007. Habitat selection and movements of raccoons on a grassland reserve managed for imperiled birds. *J. of Mamm*. 88(4):1082-1089.

Nowak, R.M. 1999. *Walker's Mammals of the World*. Sixth Edition Vols 1 and 2. The Johns Hopkins University Press, Baltimore.

Nowak, R.M. 2005. *Walker's Carnivores of the World*. The Johns Hopkins University Press, Baltimore.

Parker, K.L., M.P. Gillingham, T.A. Hanley, and C.T. Robbins.1999. Energy and protein balance of free-ranging black-tailed deer in a natural forest environment. *Wildlife Monographs* 143:1-48.

Patemaude, F. 1984. The ontogeny of behavior of free-living beavers (*Castor canadensis*). *Z. Tierpsychol.* 66:33-44.

Pearson, O.P. 1985. Predation. Pp 535-566 in *Biology of New World Microtus*. (Tamarin, R.H. Ed.) Special Publication of American Society of Mamm. 8:1-893.

Polis, G.A., M.E. Power, and G.R. Huxel. 2004. *Food webs at the landscape level*. U. of Chicago Press: Chicago, IL.

Pritchard, K. 2008. *Willow Creek vole study. A five day live-trapping project in August.*

Randa, L.A., D.M. Cooper, P.M. Meserve, and J.A. Yungar. 2009. Prey switching of sympatric canids in response to variable prey abundance. *J. of Mamm*. 90(3):594-603.

Reichman, J.J., T.G. Whittham, and G.A. Ruffner. 1982. Adaptive geometry of burrow spacing in two pocket gopher populations. *Ecology* 63:687-695.

Rezendes, P. 1993. *Tracking & the Art of Seeing: How to read animal tracks & sign*. Camden House Pub, Inc., Vermont 1-320.

Romanach, S.S., E.W. Searbloom, O.J. Reichman, W.E. Rogers, and G.N. Cameron. 2005. Effects of species, sex, age and habitat on geometry of pocket gopher foraging tunnels. *J. of Mamm*. 86(4):750-756.

Sargent, A.B. and S.H. Allen. 1989. Observed interactions between coyotes and red foxes. *J. of Mamm*. 70:631-633.

Sheehan, S.T. and C. Galindo-Leal. 1997. Identifying coast moles (*Scapanus orarius*) and Townsend's moles (*Scapanus townsen-*

dii) from tunnel and mound size. *Canadian Field Naturalist* VIII: 463-465.

Slade, N.A. and S. Crain. 2006. Impact on rodents of mowing strips in old fields of eastern Kansas. *J. of Mamm.* 69:342-352.

Tabor, J.E. and H.M. Wight. 1977. Population status of river otter in western Oregon. *J. of Wildlife Mgmt.* 41:692-699.

Taitt, M.J. and C.J. Krebs. 1983. Predation, cover, and food manipulations during a spring decline of *Microtus townsendii. J. Anim. Ecol.* 52:837-848.

Toweill, D.E. and A.G. Anthony. 1988. Coyote foods in a coniferous forest in Oregon. *J. Wildlife Mgmt.* 52(3):507-512.

Van Gelder, R.G. 1982. *Mammals of the National Parks*. The Johns Hopkins University Press, Baltimore. 1-310.

Vaughan, T.A. 1972. *Mammalogy*. W.B. Saunders Co. Philadelphia, PA. 1-463.

Verts, B.J. and L.N. Carraway. 1984. *Keys to the mammals of Oregon*. Third Edition. Oregon State University Book Stores, Inc. Corvallis, OR. 1-178.

Verts, B.J. and L.N. Carraway. 1987. *Microtus canicaudus. Mammalian Species* No. 267:1-4.

Verts, B.J. and L.N. Carraway. 1987. *Thomomys bulbivorus. Mammalian Species* No. 373:1-4.

Verts, B.J. and L.N. Carraway. 1998. *Land Mammals of Oregon*. University of California Press, Berkely, CA.

Voss, G. 2008. Personal communication based on his live-trapping of small mammals and personal observations over his years of work in Willow Creek Preserve for TNC.

Wells, M.C. and M. Bekoff. 1982. Predation by wild coyotes: behavioral and ecological analysis. *J. of Mamm.* 63(1): 118-127.

Yates, T.L. and R.J. Pederson. 1982. Moles. Pp. 37-51 in *Wild mammals of North America*. (Chapman, J.A. and G.A. Feldhamer, Eds.) The Johns Hopkins University Press, Baltimore.

Fern Ridge Lake Excursions

Paddling in our canoe atop smooth waters, my husband and I behold Nature's reflections mirroring her beautiful flora and fauna. Western grebes appear, then disappear, as they attempt to find an underwater morsel for their breakfast.

Continuing to glide effortlessly across these calm waters, we look up into the azure sky to see one of my favorite birds, the osprey, as it hovers above our heads.

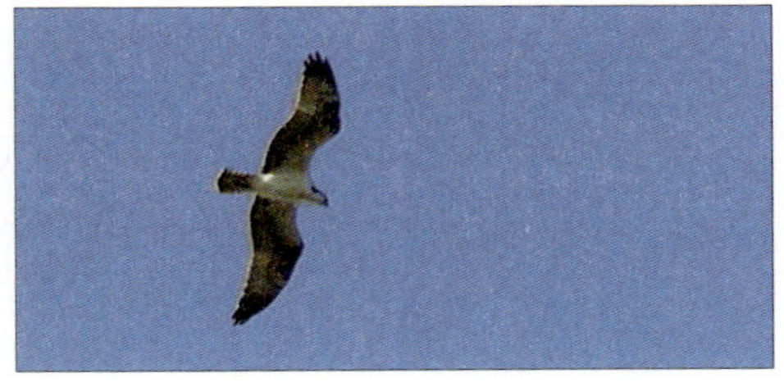

With a swift plunge into this liquid blue, it emerges with a fish in its talons. Shaking the water from its plumage, it repositions the fish head forward and flies to a nest where a female is patiently waiting. She, in turn, will share this food with their two hungry chicks. Their platform nest of sticks and other vegetative materials becomes alive with this, one of many bounties for the two growing youngsters.

I comment to the ospreys, *"Don't you think you should add a little more vegetation to your nest?"*

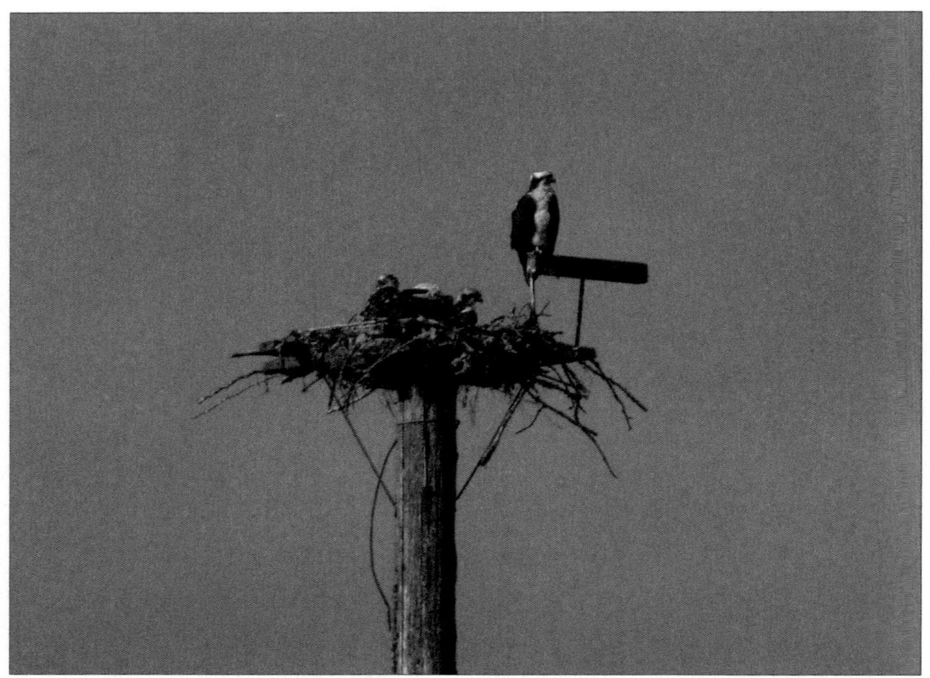

There is no response.

My concern is aroused at this point in time because the nest is rather sparsely built and the two active chicks could fall out. *Oh well, time will tell.*

Because ospreys are monogamous, they may use the same nest year after year, so maybe this is a first time for the present pair. One nest farther upstream in Coyote Creek is huge in comparison and contains an adult who is barely visible as she lies in wait for the delivery of her meal. *Are there also chicks tucked within?*

We quietly float into some grasses which hold us in their embrace so we won't disturb the large birds sitting atop the waters in a nearby tranquil setting. Five white pelicans move about so gracefully that their reflections appear like another bird lies beneath each one.

My husband slowly raises his camera for a representation to be placed in our book of memories.

The quiet waters are disrupted as the pelicans decide it is time for their breakfast. We witness their feeding behavior. While swimming, they use teamwork to drive fish

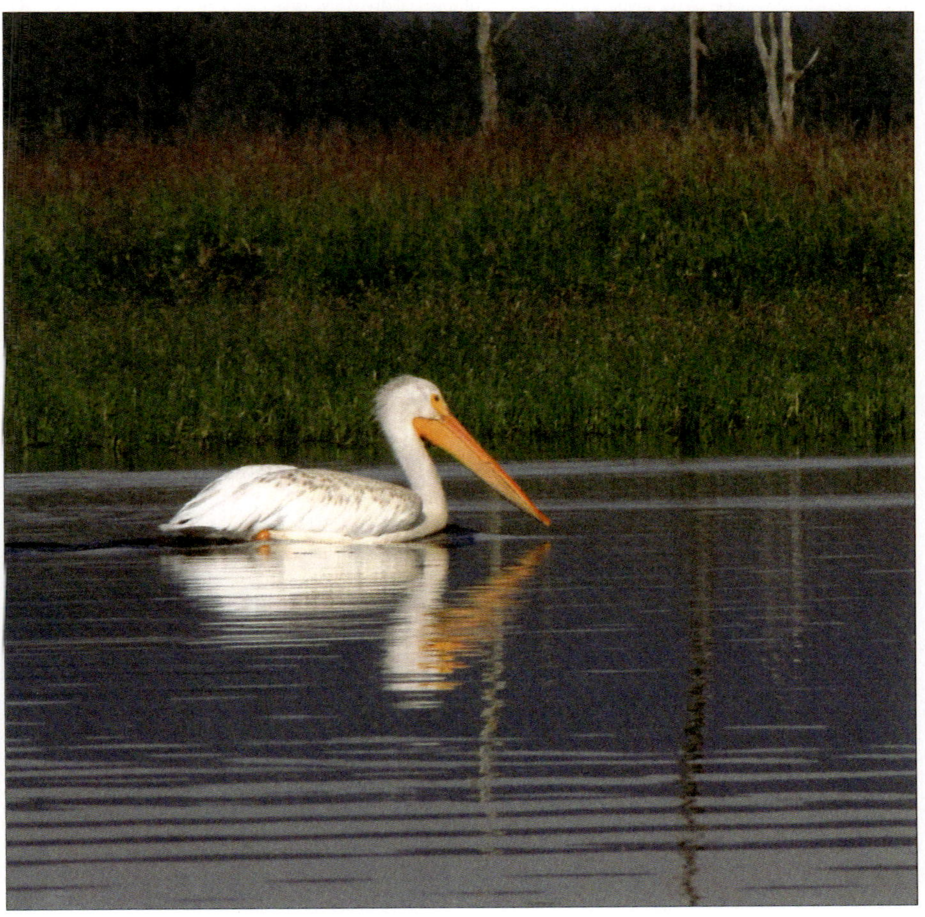

into shallow areas. Then each bird uses its long bill and pouch to scoop up the fish. This is followed by digesting their nourishment.

It is possible they have a nest nearby? It would have been placed on the ground and comprised of reeds or mud, both of which are spread about this wetland habitat. In fact, two of the birds have a gray-black crown, which indicates they are moms feeding chicks. It is thought that these crowns on nesting females help camouflage their nests from potential predators of the chicks.

Then, one by one, four of the pelicans raise their large bodies and fly like white spirits across the clear blue sky. One of the non-crowned birds remains behind, and, instead of flying, paddles along the channel of water, heading in the same direction as the others who had flown away. Perhaps it is a sentry on duty whose mission is to deter these intruding humans from the nest.

We speak softly to the pelican as he moves past us and try to assure him that we are not there to harm the adults or their potential nest, but are there just to quietly observe Nature's stunning performance. "*After all*," I speak to the pelican, "*you are one of the actors*."

Ah, is that a proud smile?

Slowly, we return to the osprey scene just in time to see the female returning with a small branch in her beak that she weaves into the nest.

"*Thanks,*" I express to her. "*That will help to keep your young safely within their wooden cradle.*"

Did she actually understand what I said to her earlier?

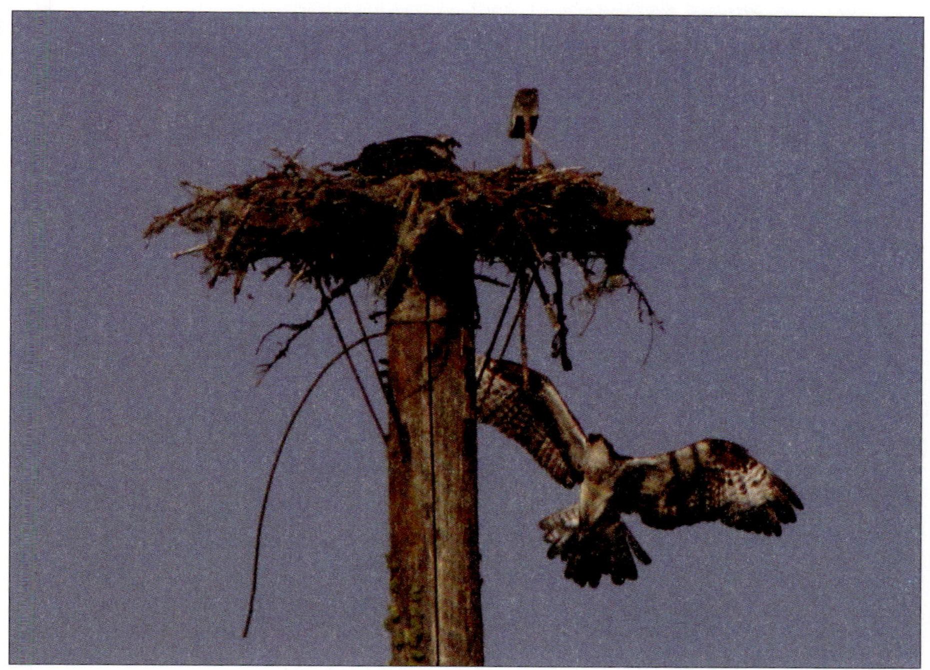

The male remains nearby on top of a tall, barren tree like a guard protecting this, one of Nature's golden treasures. I whisper "*so long for now,*" as we paddle away, and wish them continued success. This species and I connected several years ago when conducting my river otter research in the Rocky Mountains and, in a way that goes far beyond the "real" world.

Little birds twitter as they feed their own chicks tucked within small nest boxes constructed for them and spaced

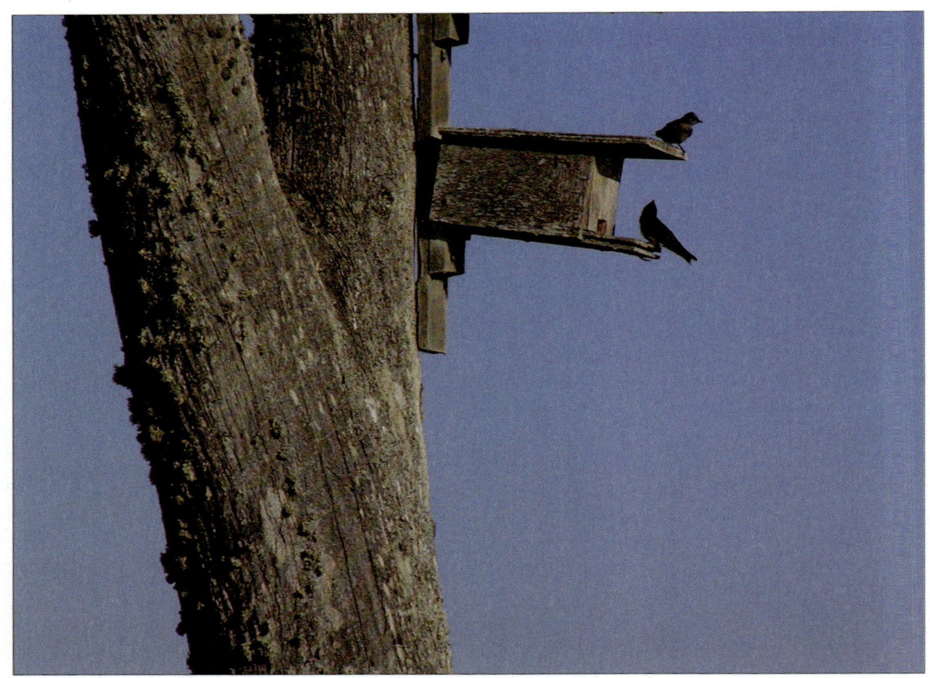

about the wetland on dead snags. As we quietly paddle through the area, avian species hidden from view vocalize their greetings (or alarms). However, the handsome birds in their black and dark-gray outfits are not at all tolerant of our presence.

Black terns, decked out in their breeding plumage, vocalize as they dive-bomb us during their air-raid attacks. "*Hey*," I yell out at them, "*we are moving away. Just give us a chance!*" Oblivious to my words, the bombardiers continue, including some white bombs dropped nearby. Juveniles perch on a nearby log — one that resembles an alligator the way it rests in the water — enjoying their parents' antics. One day they, too, will become actors on this real-life stage.

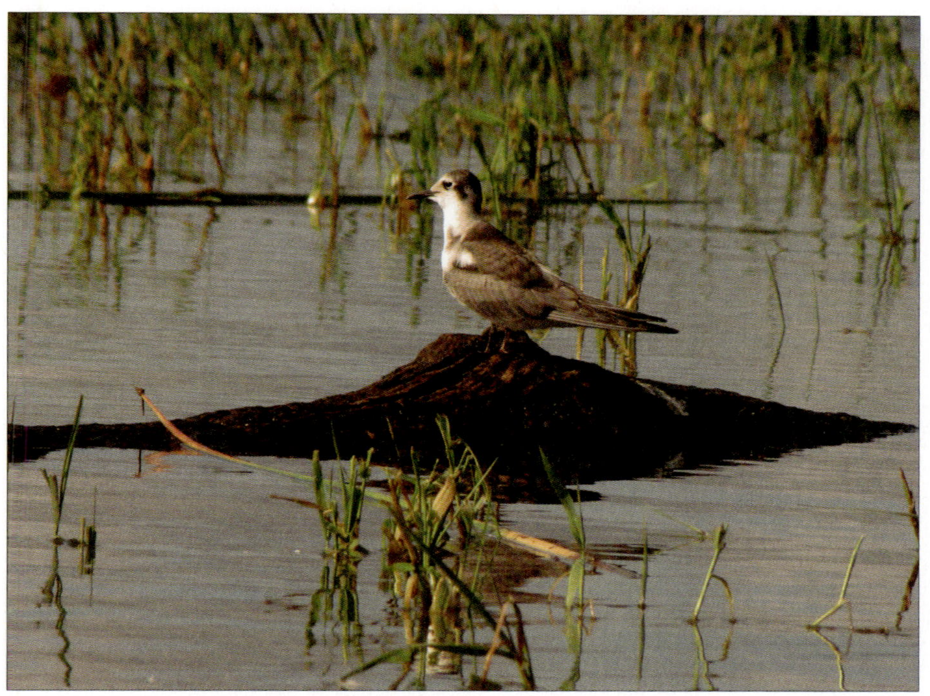

Although our intent is not to agitate them, or any of the other fauna who reside here, the species are rightfully skeptical. Far too often, the human world does not respect the homes and children of the nonhuman world.

Peace resumes as we leave the area. The western grebes continue their search for food as they paddle along, far more tolerant of our presence than were the black terns. Their swan-like necks make them appear as kings and queens of this waterway. I need not break down the whole into its parts, like in a class of science, but instead, I "feel" each ripple as the grebe dives underwater to spear a fish.

Intellectualizing this event is not necessary to appreciate its complexity. *What lies within this space of liquid form?* There is so much more than what can be seen with the naked eye. So, instead, I experience that which is occurring. Join me.

Visualize circular layers of silk spread atop these smooth waters, each layer longer than its predecessor. Enter through the center, as would a lady don her special gown, and, like the female grebe, emerge as the belle of the ball. Let the dance begin.

In fact, this species of grebe is noted for its elaborate courtship displays. With outstretched necks, the male/female pair spring to their webbed feet and, side by side, actually skip across the waters' surface. Then, at the end of this performance, they dive underwater together. Shall we try this synchronized dance? At least, we can "trip this light fantastic" through our imaginations.

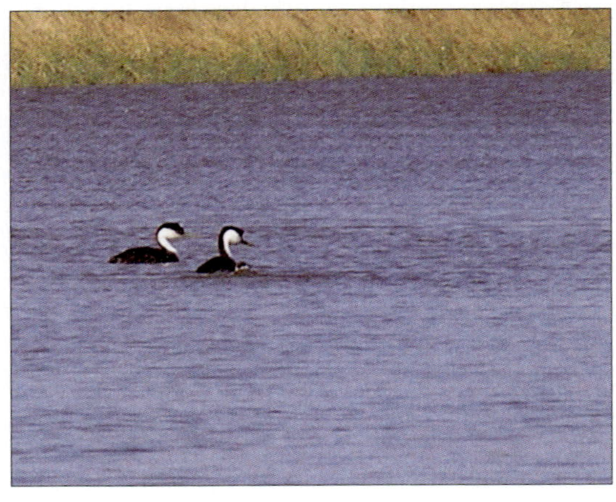

The love my husband and I feel for Nature is exemplified in this morning's outing. We can't remember such calm waters in this particular waterway, which added to the display of avian life giving a performance which brought all our senses into being. We tuck this experience into our hearts and our souls as we continue our vow to Mother Nature to do our best to respect, protect, and preserve her precious flora, fauna, and great medicine waters.

Although follow-up excursions were not on such smooth waters, we still try to continue our observations. Our light weight canoe presents some challenges, but it makes me appreciate what water-oriented birds must deal with on a daily basis. Their habitat, search for food, weather conditions, and, so many external factors are continual obstacles each day of their life cycle.

* * *

The two osprey chicks continue to grow up. Although we can observe only from afar, we see them testing their wings. What a glorious experience. One day they will take flight to seek their own territories. I can only wish them well and ask Mother Nature to guide them.

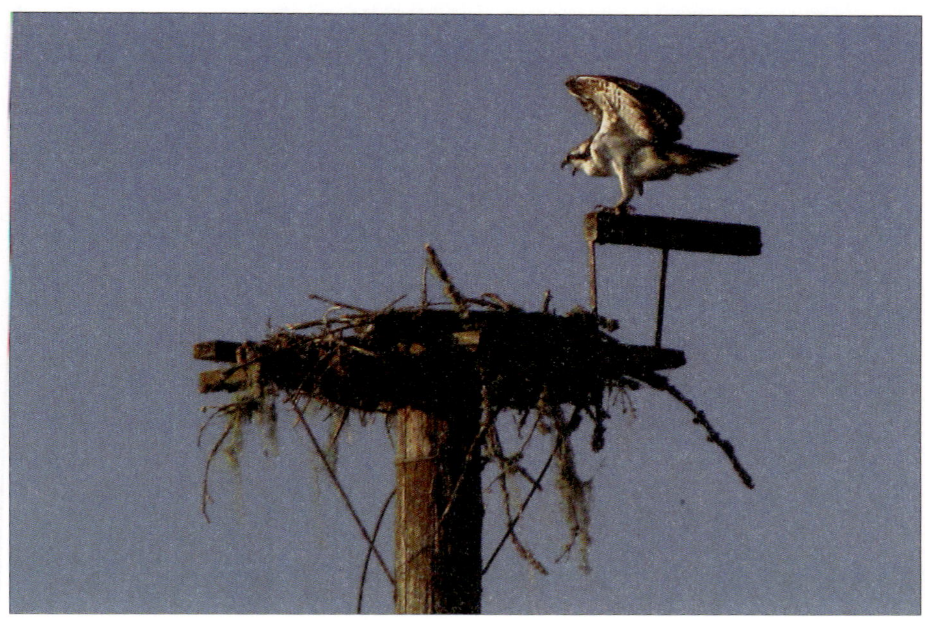

The white pelicans appear to continue their nesting behavior as we watch one of them sitting on an assumed nest. (We can't actually see their ground-level nest from our position on the water; but their behavior suggests that it's there.) If we are lucky, one day we may see their chicks.

Not to be outdone by some of the more visually appealing members of the avian world are other species of interest. We quietly enter an area in response to the short, raspy notes of the American coot, which, at one point, sounds like laughter. Their dark, slate-colored body contrasts with their high, white bill and less conspicuous white patch under their tail feathers.

During this bird's swimming behavior, we watch as it bobs its head, thrusting it forward then jerking it backward. The reason for this behavior is that during pauses between

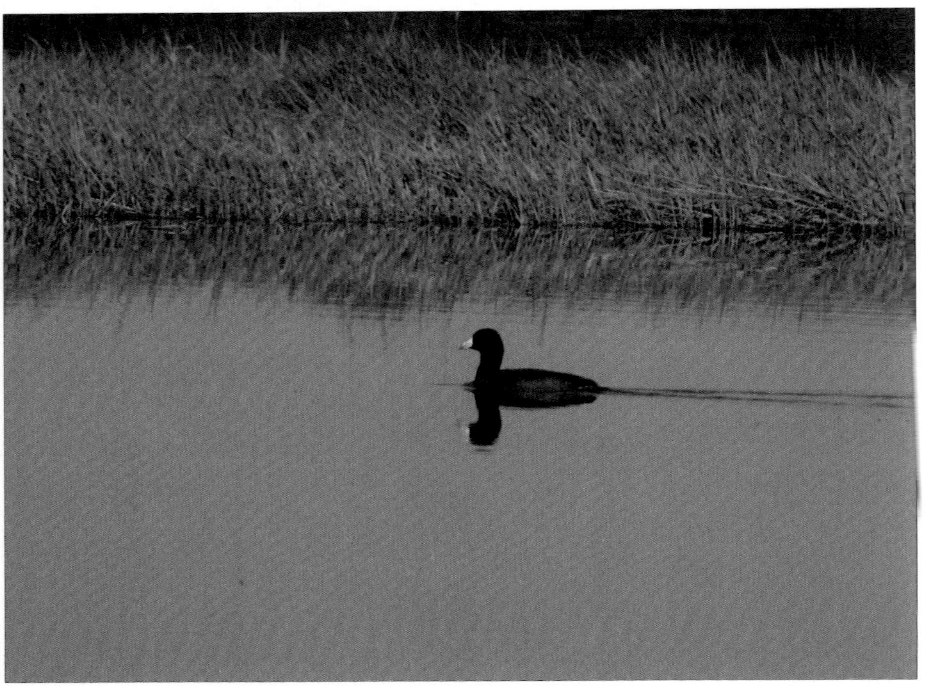

each motion, the bird can get a good lateral view of its immediate environment. This species is not a duck, but a member of the rail family.

On the stage of life, it performs at least 15 different displays, some emitted more often than others depending on the season and circumstances. One display is used during an aggressive charge, but is also used similarly during take off when leaving an area. Head positions are a clue as to whether the bird is charging or fleeing. With wings flapping and its lobed toes becoming visible above the surface, the coot actually walks on water.

Messages of the coots and all other faunal species abound during their different visual and vocal displays, which are immediately understood intra-specifically. Researchers spend years of comprehensive study trying to detect what they are communicating and their social significance.

Ah, the complexities of Nature!

Bouncing around in the light winds, with swells now rocking our canoe from starboard, we head back towards land. Along our route, we tuck in amongst some reedy islands for a break.

Here, we are greeted by the raspy call of a handsome, male yellow-headed blackbird. The bright yellow head contrasts with his black body like the petals of a sunflower radiating from its dark center. Due to its coloration, some cultures believe the male of this species is connected to a mythical being who oversees all of Nature and Nature's spirits.

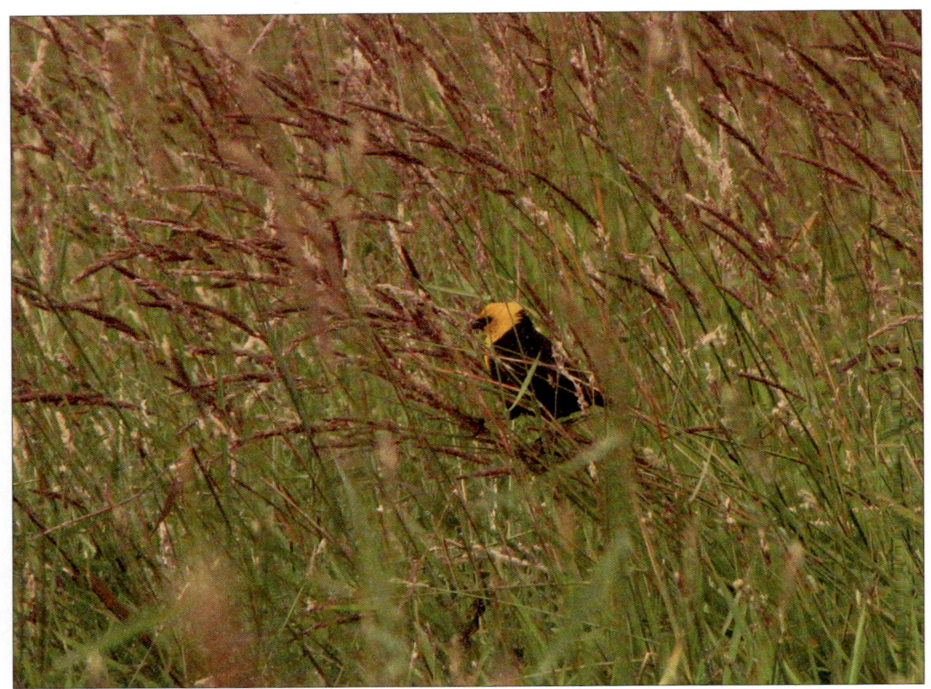

This is perfect habitat for the yellow-headed blackbird. They prefer deeper waters than their red-winged blackbird cousins, who reside in shallower waters within sedges and cattails. This corroborates our observations where both species occur within this type of habitat.

As we approach our landfall, the red-winged blackbirds sing their distinct territorial song. Not to be outdone by their cousins, the coloration of red on this species' wings with a dash of yellow is thought to give this bird ties to all creative forces of Nature.

* * *

On a different day, we paddle to the lake's confluence with Coyote Creek and head upstream in the creek. Swimming above this liquid beauty is a stunning example of a semi-aquatic mammal, the river otter. A portion of its head could be seen with its eyes, ears, and nose almost on an even plane, thus providing this species with the use of all its senses during travel through Nature's medicine waters.

This event was short-lived as the animal dove and was not seen again. There are several places for the otter to hide in this nicely vegetated land habitat. We think it entered a bank den under a large tree whose roots were outstretched arms leading towards the waters and towards the land. The otter could easily snake its long slender body within Nature's protective arms for cover.

My love of river otters has never ceased since I studied them several years ago in the Colorado Rockies. They will always be special to me, but I also know their elusive behavior makes any event of seeing just one a time of celebration.

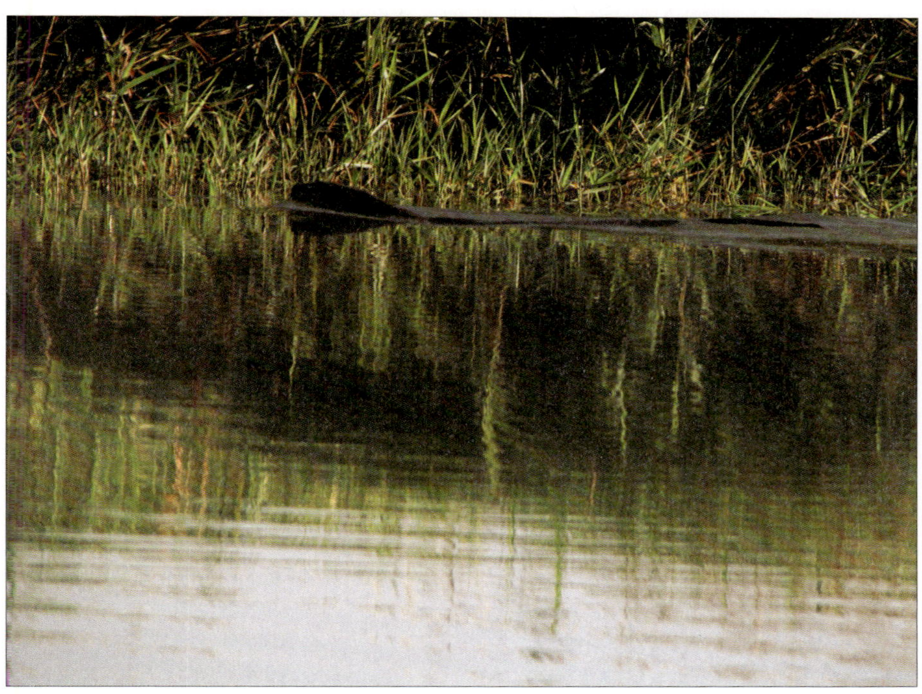

I reach out to this otter with my heart and soul. Tears fill my eyes as I imagine a photo of one of its relatives that was recently shown to me. In 2010, a deep-water oil rig in the Gulf of Mexico exploded, spilling millions of gallons of oil within its waters. This ongoing environmental catastrophic event created insurmountable destruction on the southern portions of our country. All species residing there will be affected for many years to come.

The oil entered the marshes and wetlands of these coastal states where the oiled river otter was living. I will never forget the photo of this one animal. Nor will I forget individuals photographed — and how many more that were not photographed — from far too many species. These haunting pictures in my mind are from innocent species of fauna asking, in their own way, "*Why?*"

Although just one of the many catastrophes at the hand of man, it is so far one of the worst. The real negative impacts will continue for many years and may never totally go away.

When will Homo sapiens ever learn that they share this earth with many other species, each of which is an integral part of the web of life and of which man plays a minor role? Yet, humans are the most destructive! Don't they realize their pernicious behavior is not only destroying nonhuman species but also their own?

Returning to the otter event, it is special for my husband and me as we continue to paddle upstream along quiet waters and into a "private" area. A combination of debris and assistance, provided by another semi-aquatic animal, the beaver, brought us face to face with a beaver dam. We sit silently.

And, as if on cue, a beaver appears! It is still early morning and this particular location is nicely treed and primarily free from human intervention, present company excluded. However, just like the river otter, its appearance on stage is brief. This species can stay underwater for about 15 minutes, so we patiently wait for it to surface. Being surrounded by so much lush vegetation, the minutes tick away like seconds on the clock of time.

My imagination carries me farther into the depths of this habitat. An unknown force beckons me to follow. Hugged by the arms from shades of green, their message is clear. Yes, this magnificence must always remain for both this flora and its respective fauna. Returning to reality, beaver did not reappear.

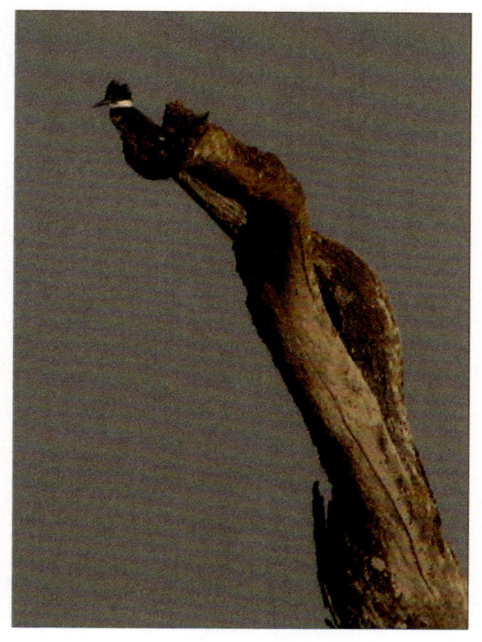

As we continue upstream, the habitat quickly transitions from wetland to riparian. The songs of unsung heroes flow from the branches of heavily vegetated trees. These songbirds are seldom seen, yet they fill the air with their harmonious sounds. We listen carefully.

Then the songsters are interrupted by a recognizable rattling call. We look up to a branch overhanging the edge of our waterway and there is a belted kingfisher.

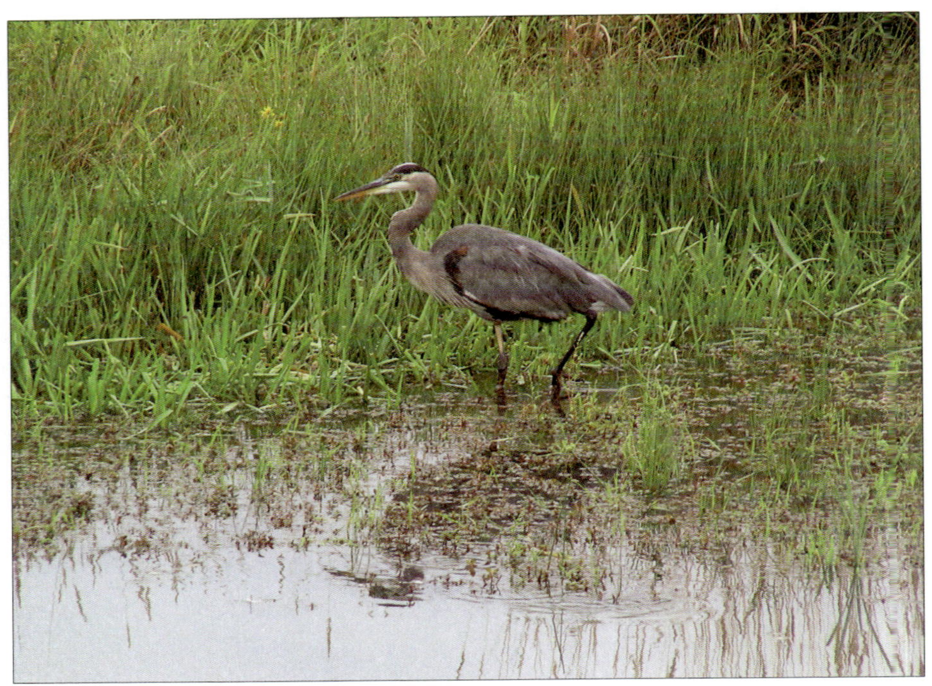

This attractive species with its blue and white markings, blue crest, and dagger-like bill is gazing into the waters for a small fish. Let's watch! There it goes diving into this liquid form headfirst, and emerges with a fish!

The bird returns to a branch and feeds on its prey. Interestingly, to some cultures, kingfishers are a symbol of peace and prosperity.

We paddle back into the lake in time to see the slow, stalking stride of a great blue heron as it stands within swallow waters. Then, with lightning speed, its sharp beak stabs a fish for breakfast. This bird is the tallest of the herons and is considered the king of the marsh. When in flight, its head is folded back in a flat S-shaped loop.

The mystical message, reflected in this great blue heron pattern, is one of an innate wisdom to be able to maneuver through life and control its life circumstances. If only this were true, then applying this maneuverability and control to saving this whole ecosystem (along with so many others) could be achieved. We caring humans must try.

* * *

It's now later in summer. Today, the overcast skies make for a pleasant outing. Because we head east to cross the lake early on summer mornings, we often face the sun for half our journey. Looking into this bright ball of light during our canoe adventures is a distraction, for sure.

Paddling along, we see movement in the water just below the surface. It looks like a large animal. *Could it possibly be a family of river otters?* We sit quietly and wait for the emergence.

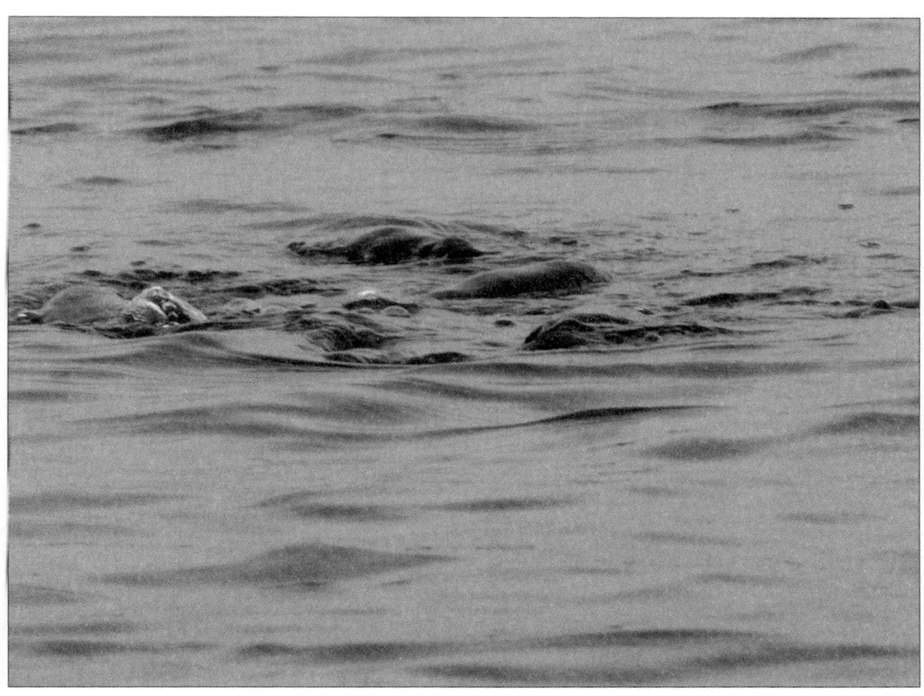

Otters don't appear but we are witnessing a behavior new to us: groups of large fish — likely carp — are swimming together with their mouths open producing a net of bubbles.

Watching this behavior makes me feel like I am sitting within an alien world. These fish do not look real, nor does their behavior. *Are they aliens from another planet?* How many times have we canoed this lake and never seen anything like this before? We've observed individual fish jumping above the surface of the water seeking their insect nourishment, but nothing like this. *Have we been transcended into the unknown?*

One group is close to our boat, but there are several others within this particular stretch of water. We even hear their sounds. *What is their message? Are we being asked to join them?* I feel uncomfortable but am mesmerized by this event.

We continue into "osprey land" and see the chicks testing their wings. Parental guidance continues for these youngsters. Otherwise, it is relatively quiet in this bird-breeding location, except for the smaller birds.

Where have the white pelicans gone? Did they have chicks? Have they already fledged? Their mystery remains unsolved.

* * *

A new day dawns over the lake. It's now September. The lake level has lowered enough to paddle under the highway and railroad bridges into an area we haven't yet explored. When the waters are higher, we can't fit underneath the two low bridges that separate this segment of the waterway from the main lake body. (It is the only canoe entrance from this location into this wide expanse of liquid form.)

While pulling ourselves along under the bridges, we are greeted by swallows who nest here. These small, insect-feeding birds help control some of the pesky insects. They are very skillful aerialists. What always amazes me is that whether under the bridge or encountering their flocks in open waters, they have never physically touched our beings in pursuit of their morsels. However, we can feel their energy.

Mother Nature presents another special gift to us today. We paddle into a new area that was unreachable in previous years. Moving through narrow passageways and over water-logged vegetation, we reach a hidden treasure.

We discover a secluded beaver lodge, nicely constructed at the water's edge. Confirmation of this home belonging to beavers is revealed by clues it left behind in the form of stripped branches lying atop the lodge with distinct tooth marks at the end.

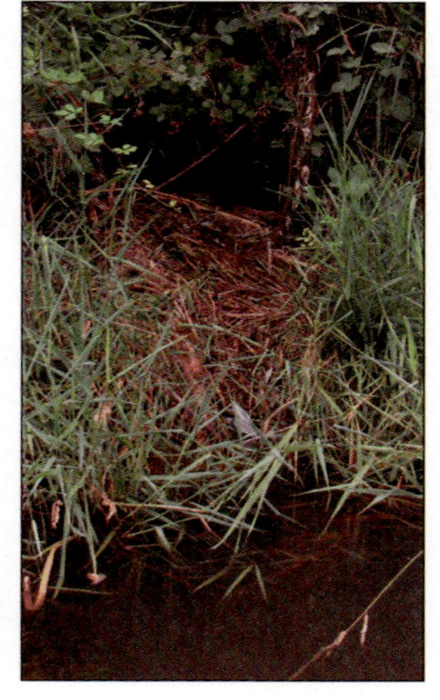

We sit quietly as sounds emerge from the area. Considering the higher frequency of the vocalization, it appears to come from a young animal. Young beavers may only be two- or three-months old at this period of time. *Have we disturbed a returning adult?*

Since we feel concerned that our sudden appearance may have caused a disruption, we continue farther along in this stretch of water. After a short period of time, we reach the end and slowly return to the lodge for further study.

All is now quiet. This "diamond in the rough" was intricately constructed by a master craftsman. There is good water depth here, determined by extending a paddle downward and not reaching the bottom. This is also due to masterful beaver engineering.

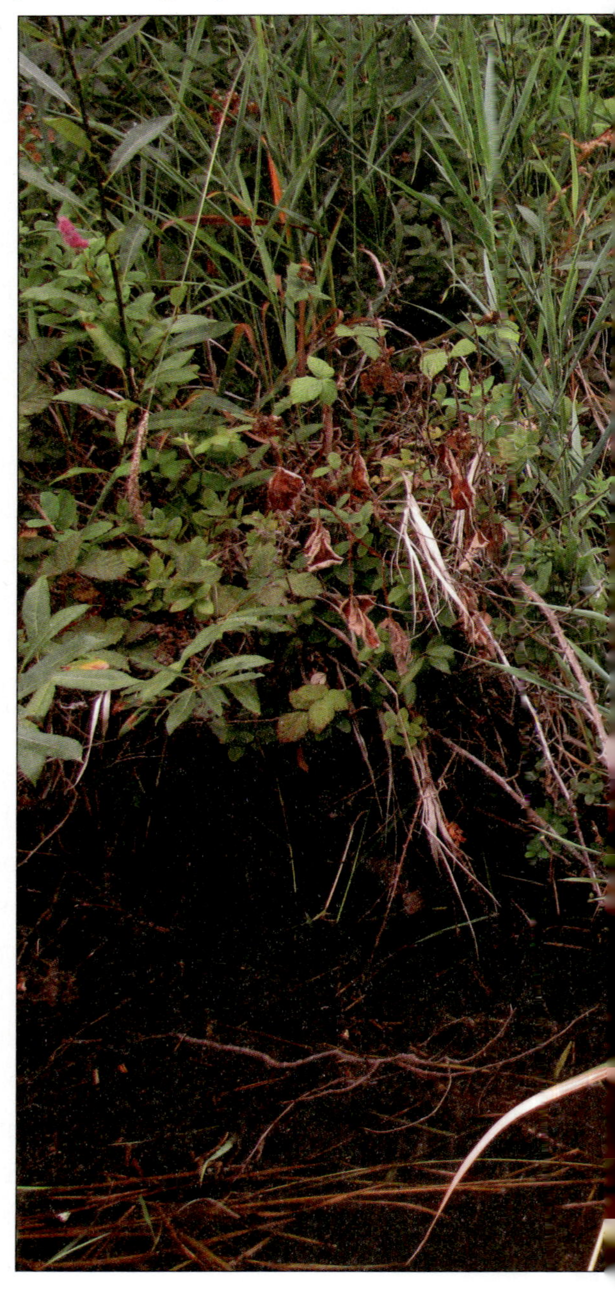

While my husband snaps pictures, I imagine the passionate eye of artist Vincent Van Gogh painting this Nature scene.

There are shades of green creatively mixed with bronze-colored branches along with golden grasses encased within the arms of Mother Nature's other floral concealment. The entrance into the lodge is underwater but next to it is an area under the roots of a tree with a distinct beaver-toothed branch extended outward. On the opposite side of this home is a vegetated slide into the water.

Please paint this Nature scene, Mr. Van Gogh. Then it could hang on the walls of this treasure. Actually, are the human artists who paint scenes like this different from the

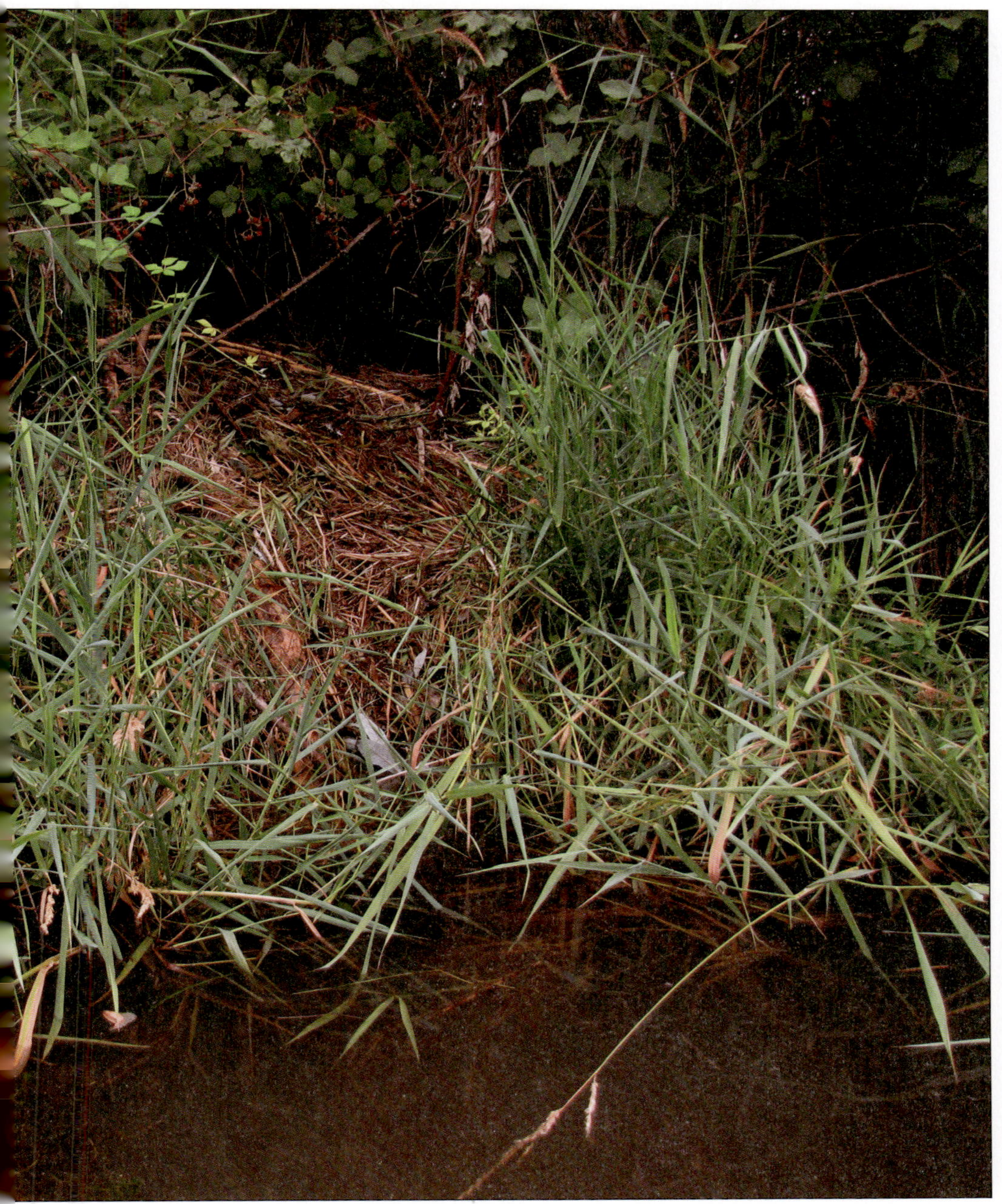

nonhuman artists who create them? Those with a love for Nature are also at one with her.

Before Van Gogh considered himself to be an artist, he wrote to his brother Theo: "Try to take as many walks as you can and keep your love of Nature, for that is the true way to learn to understand art more and more. Painters understand Nature and love her and teach us to see her."

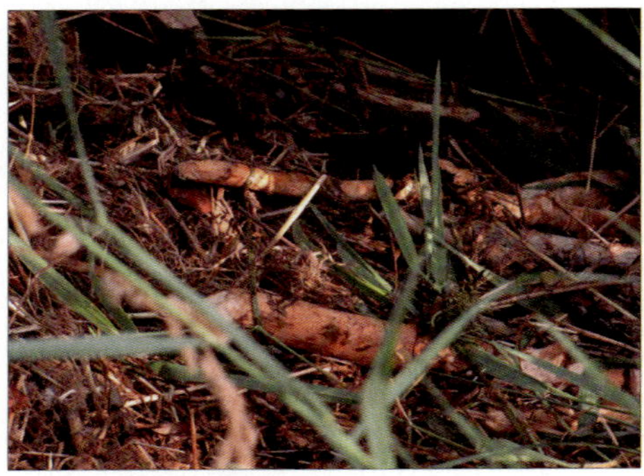

The words from this forthcoming master artist show his sensitivity for Nature, which later came through in his own art. Although there were, and still are, many excellent artists who paint scenes from Nature, I can only share with others the one who gives me personal inspiration.

Van Gogh's creations, as with those of Mother Nature, can speak directly to us all. Walk in Nature or paddle in her waters and open yourself to her with love.

* * *

The last season before summing up our canoe adventures for the writing of this journey occurred in 2011. We paddled into Fern Ridge during late spring to the unfolding of a different world. Compared to prior years, fewer birds greeted our visual senses and fewer sounds entered our aural senses from those hidden from view.

How come we don't see the fish jumping above the waters? Winter and spring both came very late to the Willamette Valley this year. *Could climate change be manifesting its effects on our wildlife paradise?*

Concern is felt for the osprey. Reaching the nest that we keep tabs on every year, we feel relief at seeing a comparatively larger nest with two adults and one chick. *Um, usually there are two chicks*. Perhaps just having one chick was this pair's response to environmental conditions. The chick appeared healthy as it stretched its body up from the floor of the platform nest. *That's good.* We observe quietly for awhile. Then one of the parents flies away.

Paddling onward in this area, we find the white pelicans nicely settled in on tranquil waters. At least, they provide

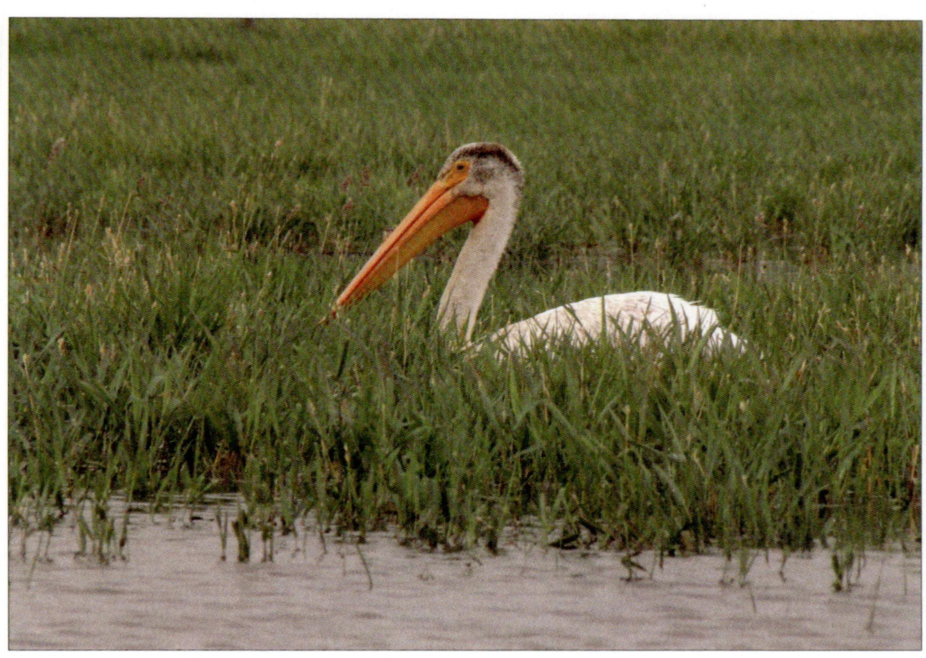

some comparative consistency. But fewer western grebes greet us compared to prior years. The haunting sounds from the hidden pied-billed grebes are scattered throughout the marsh along with a few calls from species of mystery. Again, not in the number of sounds compared to years past. Continuing to the very large osprey nest, no birds are seen. This creative work of Nature is sitting like a beacon that not all is right with the world.

At least, some of the smaller species which use nest boxes provided for them have returned, along with a few of the yellow-headed and red-winged blackbirds. *But, where are the American coots and the great-blue herons?*

Because our canoe outings are for enjoying the beauties of Nature, we have not applied the scientific method to our observations. However, it is plainly obvious that what we are currently viewing is quite different than what we observed during this time period in prior years. It has been a cool, wet spring, which could be affecting the Natural World, so we'll see what transpires in the coming weeks. We hope that things have just been delayed this year.

We return to the occupied osprey nest just in time to see an adult bring a fish to the anxiously awaiting youngster. "*Thank-you, Mother Nature, for allowing us to witness this one special event,*" I speak to her. "*Please continue to watch over them and all bird species still surviving in this one small segment of the Natural World.*"

* * *

There is one area of this large lake that we visit at least once a year, but since we see fewer species of fauna, it is not a priority. Another reason has become evident today:

winds toss us around in light-hearted play. *"Hey, I don't appreciate your type of play,"* I speak to the winds.

Happily, we return to shore and just in time to be greeted by a first for us. A mink suddenly surfaces and swims towards shore, moving up onto a log at the edge of the waterway. But as quickly as it appears, it disappears, moving onto land and into a heavily vegetated shoreline. We quietly wait in hopes of the animal's return, but no such luck. At least, we received a pleasant surprise. So this time, paddling in this location was worth the effort.

* * *

We paddle into mid-summer and the eventual return to some semblance of normalcy for species in the lake's wetlands. *Is this the beginning of a new trend or just a spike on the continuum?*

Coots are back displaying their behavioral repertoire; more sounds emerge from those species hidden from view. *Ah*, there is a great-blue heron which appears to be conversing with a large white egret. We can guess at that parley. The heron is the first to fly away, while the egret remains. *Ummmm.*

Returning to our osprey nest, the osprey chick matured and eventually fledged. What we believe to be the former chick remains in the area of its parents. I only wish this majestic bird could answer some of our questions about that which is occurring. Vocalizing, perhaps it is trying to reply.

Then, in mid-August, we receive another surprise. A new chick is observed in the nest. The female double-clutched this year, but with one chick each time. Species of the Natural World respond accordingly to ambient conditions.

We only hope this chick will have time to fledge before other factors take over, such as lowering the lake for the upcoming rainy season.

Beautifully perched on two snags are three common egrets. One can only guess at their potential composition.

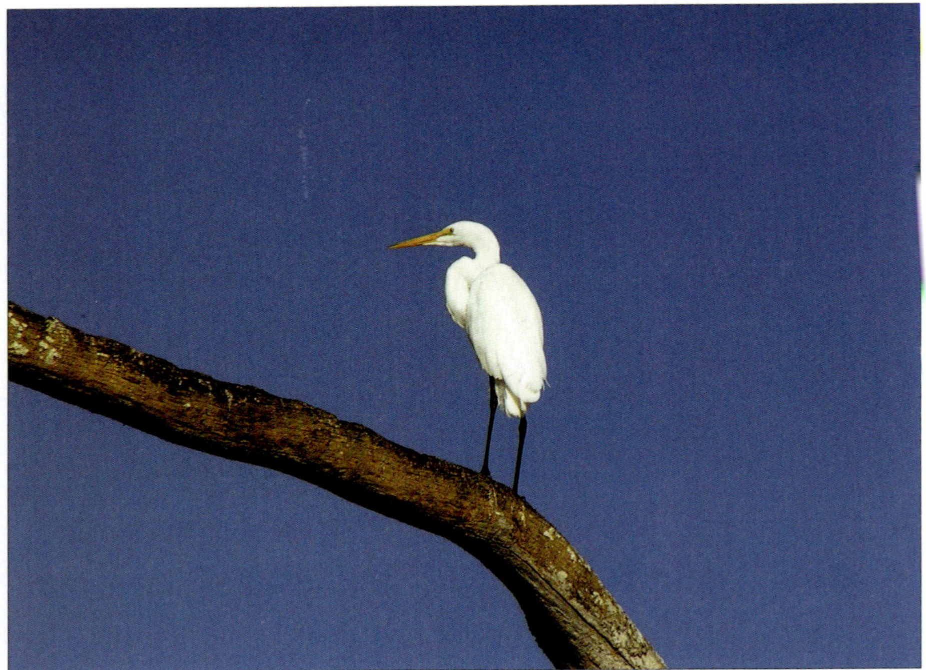

Whatever the case, their purity of white is striking against the deep blue of the sky.

Not to be outdone by the egrets are white pelicans in flight. At first, there were two, but soon they were joined by six others. Their spirited beauty spreads across the sky.

When the waters were low enough so we could get under the bridges, we return to the beaver bank lodge. All is quiet but there is evidence of new activity, with some small branches placed near the lodge. Also, some stripped wood cuttings are worked into the construction for reinforcement. All indications are that someone is living here.

We continue beyond the beaver masterpiece and into another new area for us. Actually, we pull our way along by grasping onto tangles of branches. We feel like explorers moving through an unknown land. What mysteries hide within this maze of vegetation?

Before returning to shore, we see a young western grebe

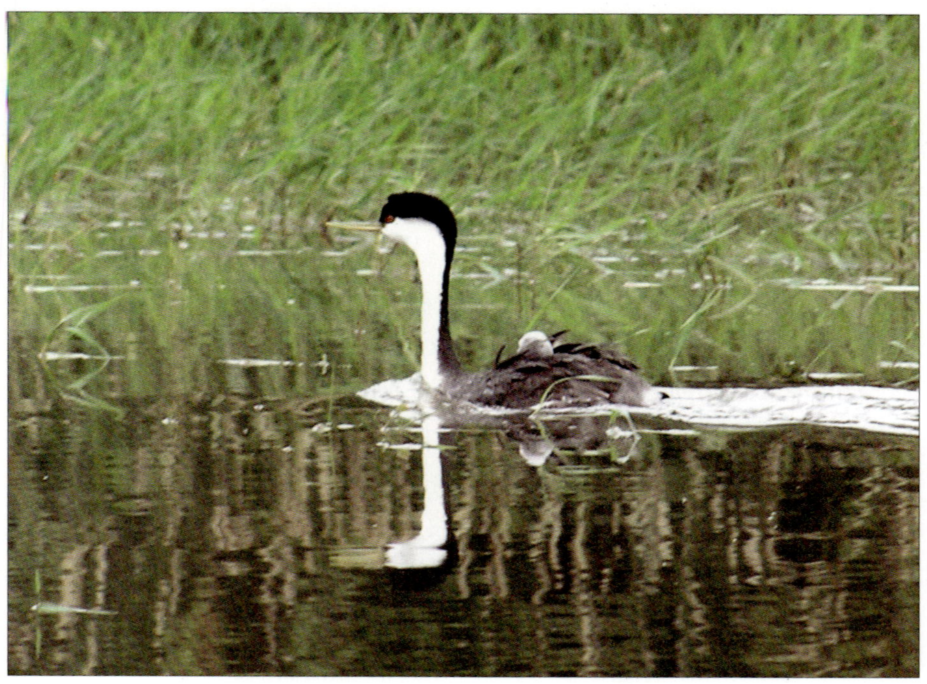

riding up on its mother's back. That is always a special sight. *Hold on, little one.*

Farther along, a different chick is trying to keep up with its parent, looking like a child's small wind-up toy.

Our last outing for the season occurs in late September and to incredible ambient conditions. Initially, a light fog immerses us into a sense of mystery. This atmospheric condition plays with unknown events on the other side. We slowly paddle along smooth waters and soon witness the gradual lifting of the fog into bright sunny skies. Fortunately, there are no winds.

Moving into the area of our osprey, a young bird is observed calling from the nest. An adult perched near by appears oblivious to the youngster. *Is fledging being encouraged? This is tough love at its best.* Farther along, we see another large osprey perched on top of a tree. This is the location where the assumed first fledged chick was observed. *Oh, so many unanswered questions remain for the scenario of this one family of osprey.*

Continuing our paddle, I feel like we are receiving a so-long-for-now salutation by the osprey, along with many other species greeting us as we move through the end of our seasonal journey. Over areas of open land, we watch a marsh hawk (now called a Northern harrier) gliding just above the ground for small prey, such as voles. There is a great-blue heron looking for its breakfast. Common egrets appear along our route along with western grebes and "laughing" American coots.

Then an amazing special mammal sighting occurs. I feel like I am viewing a sculpture of the species since it lay motionless on top of a log and is such a perfect representative. (In case you haven't guessed, it is a river otter!)

We never expect to see any, so we aren't paddling as quietly as we should be. The animal's response is to slide into the water during our approach. At least, my husband takes a couple of photos for our book of memories before its disappearance into the unknown. *Wow!*

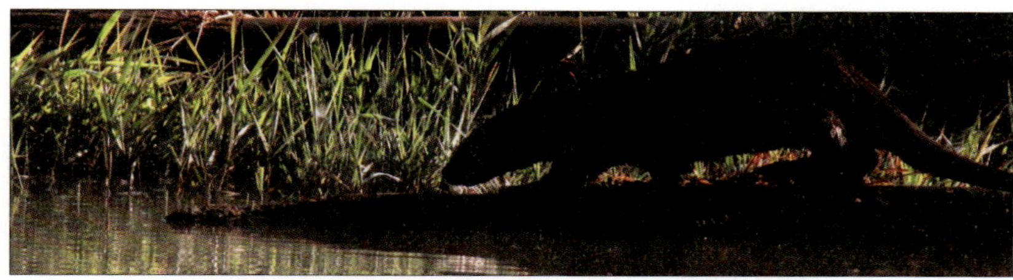

Even though we don't want this last day of the season to end, we know that, come next May, we will return to new journeys in Fern Ridge Lake. In the meantime, all the special experiences we have here will be embedded in our cycle of life.

There are many species of fauna that exist in this one relatively small segment of the Natural World. I have only given a sampling of some of the bird and mammal species that reside where we like to paddle. Now, many more of us should explore and see what else we can discover on our journeys.

* * *

The Natural World is full of mysteries and surprises for those who take the time to use their senses to explore this world. Maybe if more Homo sapiens look and feel beyond their own species, all species could have a more peaceful existence. After all, the layered complexities of floral and faunal species of Nature are each an integral part of the web of life, and thus, the existence of earth.

Whether we walk in Nature or paddle in her great medicine waters, we must learn to respect, protect, and preserve the Natural World. Mother Nature keeps trying to communicate this to us humans. Are we among those who are listening? For the sake of all living existence, I hope so!

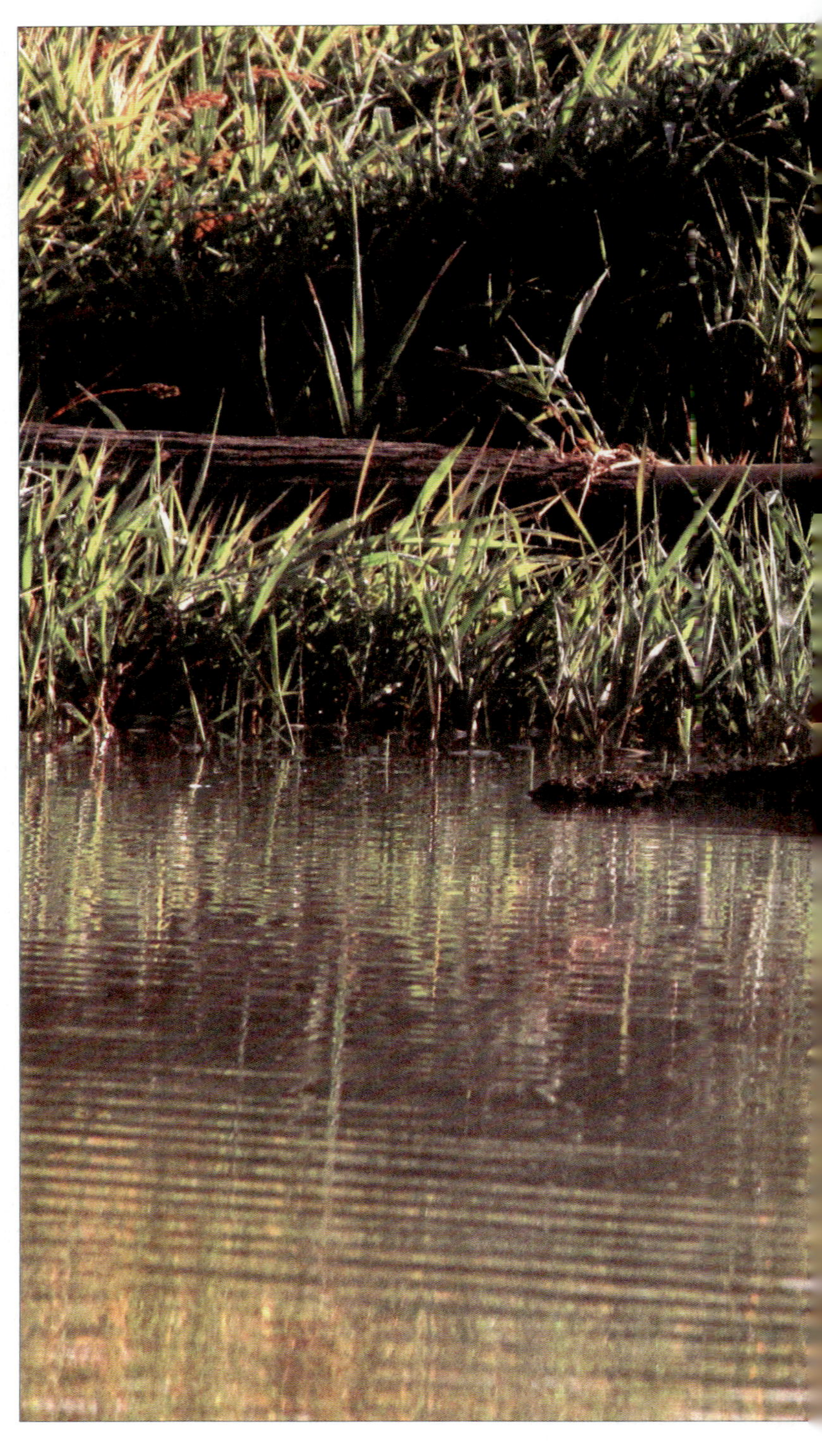

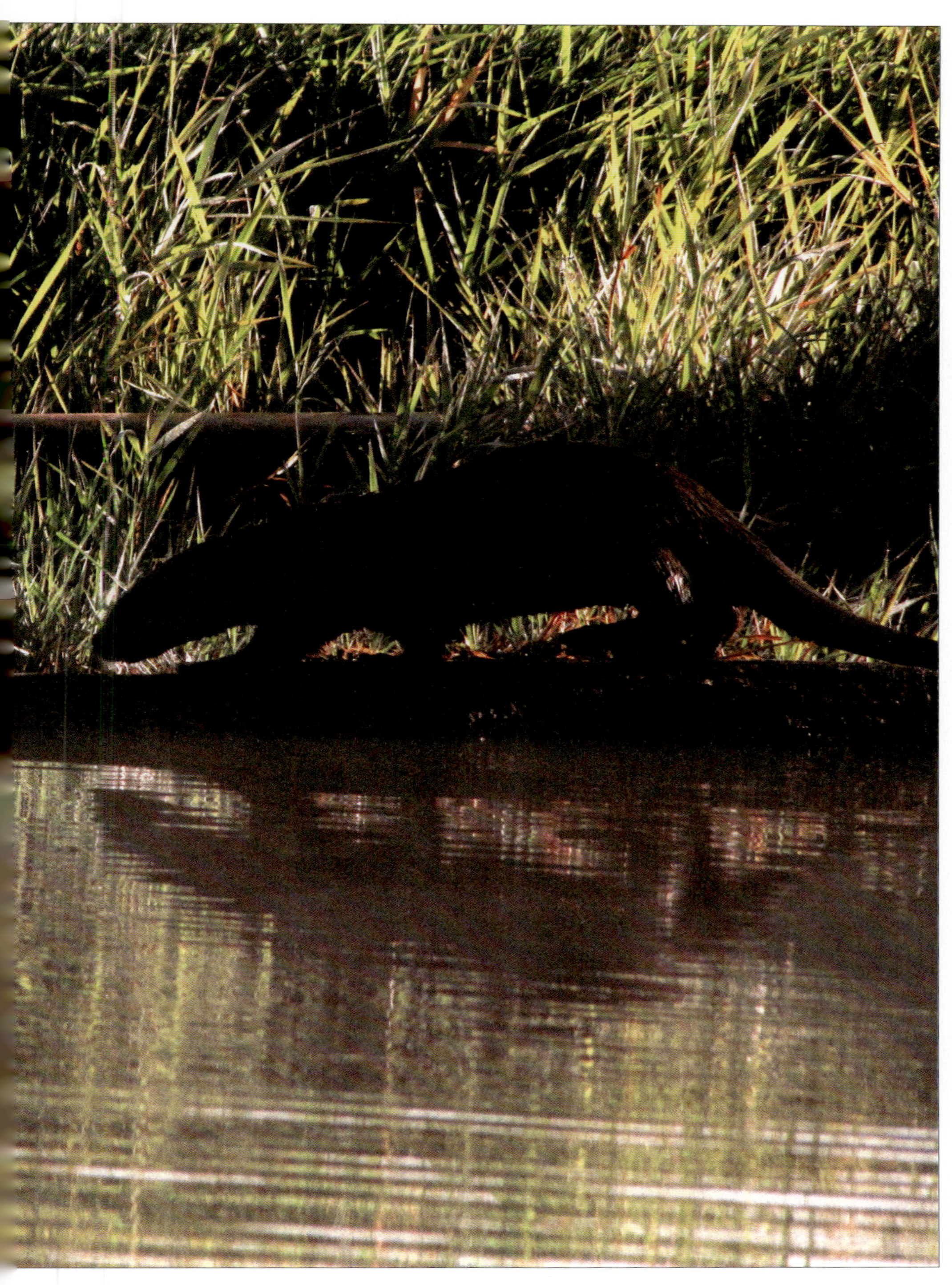

Reference

Comments on powers associated with some of the bird species in Fern Ridge Lake Excursions:

Andrews, Ted. 1997. *Animal-Speak: The spiritual and magical powers of creatures great and small*. Llewellyn Pub., St, Paul, MN.

Post Script

The endearing beaver must always be
clever recycler of Nature's tree.
They build their dams which help another
through evolution from the Earth's Mother.

This wetland-riparian ecosystem engineer of the Natural World continues to amaze us. Our journey with the beaver portrayed their natural history, while illustrating some of the many advantages attributed to the work of this one animal. Their impact on Nature benefits many floral and faunal species, including Homo sapiens.

Whether it is drought-stricken areas or rain-drenched locales, if allowed, the beaver can assist with the problem. Particularly in the western part of our country, from ranchers to stream conservationists, beavers have been and continue to be called upon to "lend a helping hand."

I remember reading an article in *National Wildlife*'s April/May 1994 edition (pp. 24-27) entitled "Bullish on Beavers" by Stephen Stuebner. This article described how ranchers in Colorado, Idaho, Montana, Nevada, Oregon, and Wyoming are working with federal officials to restore depleted riparian habitat by bringing in beavers. Because cattle overgrazed and depleted ranchlands throughout the West, beginning with the European invasion in the 1800s, streams became silted over from soil erosion, flooding the ranches during rainstorms and delivering impurities downstream. Finally, at least by the 1970s, some ranchers began turning the tide to maintain stream integrity with help from a very willing worker, the beaver. They didn't even ask for a paycheck.

Articles continue to appear in conservation magazines and other publications to make us aware of how the work from this one species is helping repair some of the human and livestock degradation of our precious waterways.

When the Zuni Indian Reservation in New Mexico released beavers into degraded water sheds, eventually, more riparian vegetation grew in and many species of fauna moved into the improved habitat, including the endangered willow flycatcher. Researchers in Oregon, Alaska, and British Columbia discovered that juvenile salmon have higher survival rates in streams with beaver ponds. Oh, did I mention the re-introduction of beavers into the San Pedro River of arid southeastern Arizona? The beavers not only improved the habitat for other fauna, but their ponds allowed water to seep into the underground aquifer of these drought stricken lands rather than just run off.

The list goes on and on as more people learn about the benefits gained through the work of beavers. A recent article in *Audubon Magazine*'s Sept/Oct 2011 by M. Nijhuis (p. 20) stated: "Once maligned as pests, beavers are proving indispensable." Researchers continue to discover benefits bestowed by the work of this one species to watersheds throughout our country. Even the negative aspects of acid rain are turned around through beaver ponds. While beavers do what comes naturally, the complexity of science related to benefits from their behavior continues to be discovered. In fact, as time moves onward, beavers may become even more important by helping us solve many of the problems associated with climate change as it affects shifts in wet and dry seasons and wet and dry regions.

Beavers were once brought to the edge of extinction by humans. Could it now be possible that people will learn to work with and co-exist with them? Let us hope that is the case. Because if there are any problems related to beaver behavior, there are many non-lethal methods of handling them.

I end with a quote by two long-term beaver researchers, B.A. Schulte and D. Muller-Schwarze (p. 122): "Given the flexibility of beaver behavior, perhaps we would be better to manage human activity, to use preventative measures to avoid problems with beavers, and to reap the benefits of living with beavers."

I ask you to please "TRY."

Reference

B.A. Schulte and D. Muller-Schwarze. 1999. Understanding North American Beaver behavior as an aid to management. In *Beaver Protection, Management, and Utilization in Europe and North American Beavers*. (Eds. P.E. Busher and R.J. Dzieciolowski.) Pp. 109-128. Kluwer Academic/Plenum Publishers, New York.

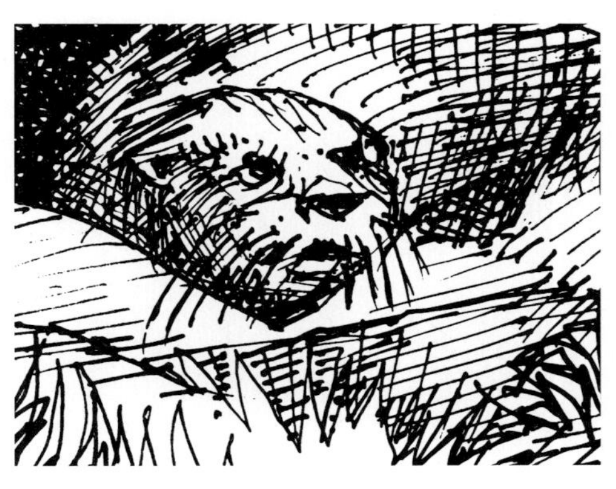

Also by

Judith K. Berg . . .

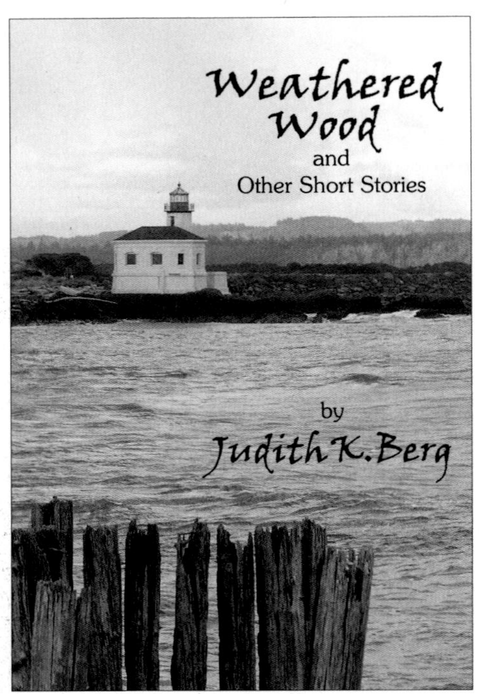

Weathered Wood

Paperback ISBN: 978-1930580909
Eight short stories, 96 pages.

Judy explains how the stories originated:

"Wanderings of my mind guided my creative inspiration into exploration of the short story. Many ideas for these stories materialized during the long hours spent as a professional wildlife researcher alone within the Natural World. Others sprang into my conscience during vacation breaks from my work, while writing *The Otter Spirit*, and just randomly from unknown corners of my mind. What emerged from this journey are stories of mystery, humor, and our finite existence. The stories are all fictional; the characters are all products of my imagination."

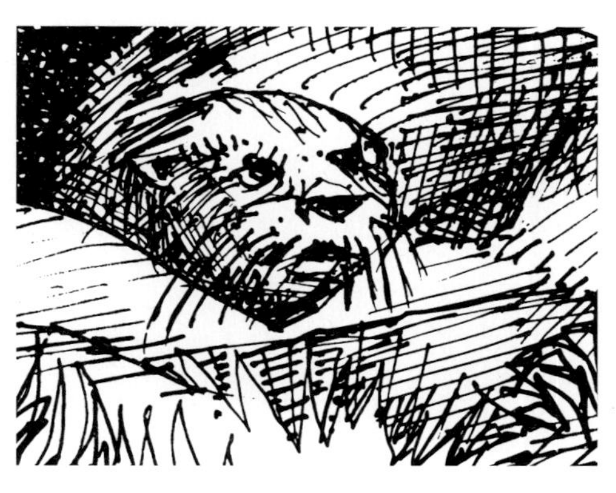

Also by

Judith K. Berg . . .

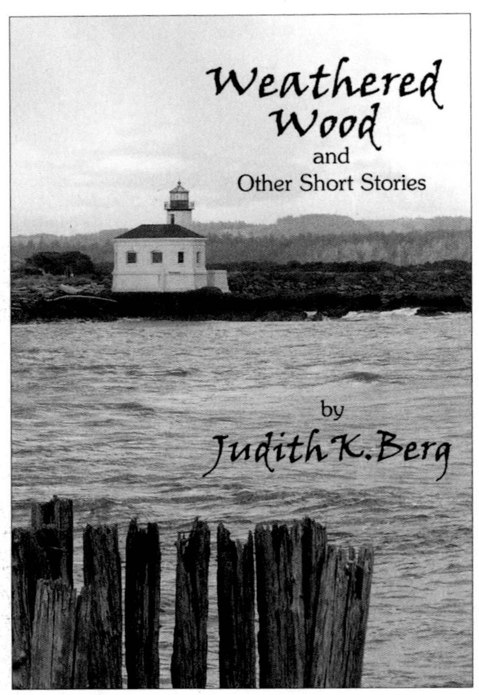

Weathered Wood

Paperback ISBN: 978-1930580909
Eight short stories, 96 pages.

Judy explains how the stories originated:

"Wanderings of my mind guided my creative inspiration into exploration of the short story. Many ideas for these stories materialized during the long hours spent as a professional wildlife researcher alone within the Natural World. Others sprang into my conscience during vacation breaks from my work, while writing *The Otter Spirit*, and just randomly from unknown corners of my mind. What emerged from this journey are stories of mystery, humor, and our finite existence. The stories are all fictional; the characters are all products of my imagination."

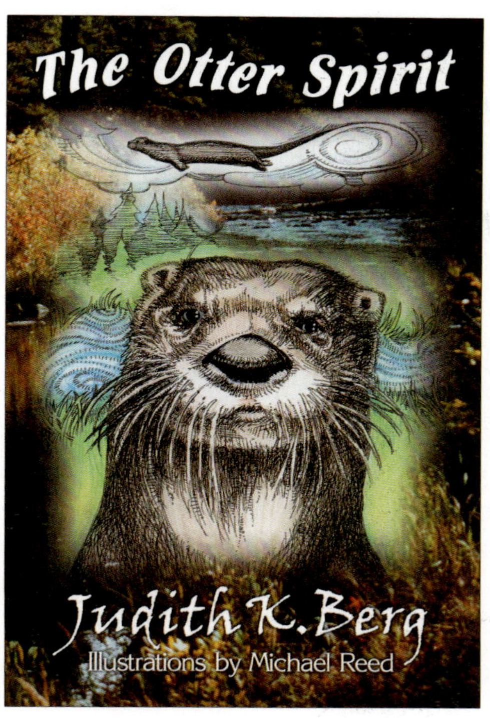

The Otter Spirit

Hard Cover ISBN: 978-1930580701
Illustrations by Michael Reed. 184 pages with references and index.
Timeless creative non-fiction for all readers.

Winner of the 2005 NABE Pinnacle Book Achievement Award.
Runner-Up in the 2009 Green Book Festival Awards.
Honorable Mention in the 2010 Eric Hoffer Awards.

Follow award-winning wildlife researcher Judy Berg on the life journey of a North American River Otter. Reflect on her philosophical introspection of the Natural World and its connection to Native American spirituality. Listen to the ancient voices and realize the sacred gifts that Nature has to offer.

Descriptive poems, interpretive illustrations, researched facts, personal photos, Native American folklore, firsthand observation and narration blend together and elevate this natural history to a class of its own.

Visit us at www.otterspirit.org

Comments about *The Otter Spirit*:

"Vivid and alive characterization of otters, their biology, their habitat, and their behavior paint clear images that take the reader into the field contribute to an informative and entertaining story."

—Dan Salzer, The Nature Conservancy

"Smooth and easy reading that conveys substantial information about the habits of various animals, but maintains a healthy distance from the scientific vocabulary that could easily slow down its delivery. Senses are stirred in the flow through the seasons, providing that necessary feeling of the passage of time equating to endless."

—Renny Rapier, Writer and Editor

"I love the idea of merging two journeys—the physical one with Nature and tracking the otters—with the spiritual one, which involves the communing with Nature that Native Americans cherish."

—*Writer's Digest*

"Appreciation for this sacred earth, from being out exposed to Nature, shows in every line."

—Mark Andrew, Sculptor

"Researcher Berg's love of wildlife and nature leads her to contribute information that will inspire others to care about and want to preserve the otter species. We learn surprising and interesting facts that go beyond the basics. *The Otter Spirit* is a unique and lyrical interplay of observations and imaginations. But the imaginations are anchored in education."

—Kathye Fetsko Petrie, *The U.S. Review*

Visit us at www.otterspirit.org